Library of
Davidson College

FRICTION

FRICTION
An Introduction to Tribology

Frank Philip Bowden
and
David Tabor

ANCHOR PRESS/DOUBLEDAY
GARDEN CITY, NEW YORK
1973

ISBN: 0-385-05109-3
Library of Congress Catalog Number 72–84969

Copyright © 1973 by Doubleday & Company, Inc.
All Rights Reserved

Printed in the United States of America
First Edition

THE SCIENCE STUDY SERIES

THE SCIENCE STUDY SERIES originated as part of the Physical Science Study Committee (of Educational Services Incorporated, now Educational Development Center) program for the teaching and study of physics. The series offers to students and to the general public the writing of distinguished authors on the most stirring and fundamental topics of science, from the smallest known particles to the whole universe. Some of the books tell of the role of science in the world of man, his technology and civilization. Others are biographical in nature, telling the fascinating stories of the great discoverers and their discoveries. All the authors have been selected both for expertness in the fields they discuss and for ability to communicate their special knowledge and their own views in an interesting way. The primary purpose of these books is to provide a survey within the grasp of the young student or the layman. Many of the books, it is hoped, will encourage the reader to make his own investigations of natural phenomena.

ACKNOWLEDGMENTS

The first draft of this book was completed shortly before Professor Bowden's death. Because the publishers felt it assumed too great a scientific knowledge on the part of the general reader I have rewritten it. I have, however, retained with gratitude and pride the name of Professor Bowden as co-author: for his contribution to this book is very real, in spirit if not in words. If any individual brought science into tribology it was surely Philip Bowden.

I wish to express my thanks to Miss Jennifer Blount for her patience in typing both the first and second manuscript, to Mrs. Carol Brown for preparing the figures and checking them and to my sister Miss Esther Tabor for making a number of helpful comments on several of the revised chapters.

David Tabor

Surface Physics
Cavendish Laboratory
Cambridge, U.K.

December 1971

Frank Philip Bowden, 1903–68

FRANK PHILIP BOWDEN was born in 1903 in Tasmania and educated there at the university in Hobart. He went as a Commonwealth Scholar to Cambridge in 1926 and apart from the war years spent the rest of his working life there. In September 1939 while on a visit to Australia the war broke out and at the request of the Australian Government he established a laboratory (later known as the Division of Tribophysics) to deal with tribological problems connected with the Australian war effort. On returning to England he re-established his research group there and extended his interests into other fields. He was also active on government committees and his advice was widely sought in industry. He was a Fellow of Gonville & Caius College, Cambridge, and was elected a Fellow of the Royal Society in 1948.

Philip Bowden was an experimental scientist of great originality, perception and versatility. He made fundamental and pioneering contributions in the fields of friction, explosive sensitivity and fracture and he won worldwide recognition in all these areas of research. He was also a man of great charm with a keen appreciation of the aesthetic side of life and the influence of this quality is felt in all his scientific work. He had many interests outside science and was involved in many activities in the world of affairs. But his deepest and most sustained interest throughout his life was his laboratory and the challenge and excitement of scientific research.

David Tabor

DR. DAVID TABOR was born in London and educated at London and Cambridge universities. He worked in Australia during World War II. He was International Fellow at Stanford (1956), UNESCO Visiting Professor, Jerusalem (1961), Russell S. Springer Visiting Professor, Berkeley (spring 1970). He is a Fellow of Gonville and Caius College, Cambridge, a Fellow of the Royal Society and currently head of Surface Physics at the Cavendish Laboratory, Cambridge.

CONTENTS

I   FRICTION IN EVERYDAY LIFE   1
What Is Friction?—The Nuisance of Friction—The Desirability of Friction—How We Define Friction: The Laws of Friction—What Causes Friction? A New Word: Tribology

II   FRICTION IN HISTORY. EARLY IDEAS—GOOD AND BAD   9
The First Ideas—Academies of Science and the Industrial Revolution—Amontons and the Early French Workers—Desaguliers' Ideas of Adhesion in Friction—Coulomb's Classical Study: The Importance of Surface Roughness—How Is Energy Dissipated?—Surface Roughnesses and the Area of Contact

III   SOME PROPERTIES OF SOLIDS—AND OF LIQUIDS   25
Why Is a Solid a Solid?—The Size of Atoms: How to Sketch the Forces Between Them—Elastic Deformation—Plastic Deformation—Fracture—Brittle Solids—The Crystalline Structure of Solids—The Free Surface—The Amorphous Structure of Solids—The Structure of Rubber—The Structure of Plastics or Polymers—The Structure of Wood—The Liquid State—The Gaseous State

IV   WHAT IS A SOLID SURFACE AND WHAT HAPPENS WHEN TWO SURFACES ARE PUT TOGETHER?   47
The Importance of Surface Contours—Stylus Method: The Profilometer—Oblique or Taper Sectioning—Optical Methods—Electron Microscope—The Shape of Surface Rough-

nesses—The Contact Between Surfaces—The Nature of Metallic Surfaces Used in Engineering

V FRICTION AND ADHESION OF METALS   61
Adhesion and Plowing in Friction—The Evidence for Friction Junctions—Adhesion of Metals—Adhesion and Friction—The Wringing of Block Gauges—Static Friction and Kinetic Friction—Stick-slip Motion—The Effect of Hardness on the Friction of Metals—How to Cheat the Laws of Friction—Bearing Alloys—Some Typical Friction Values

VI FRICTION OF NON-METALS   77
Polymers and Plastics—Effect of Load and Geometry on the Friction of Polymers—The Friction of Teflon or Fluon—Uses of PTFE—Friction of Rubber—Friction of Mica—Friction of Wood—Brittle Solids—The Friction of Diamond and Other Hard Solids—Friction of Ice and Snow—Friction of Layerlike (Lamellar) Solids

VII FRICTION UNDER EXTREME CONDITIONS   95
Friction in High Vacuum: Surface Cleaning by Heating—Junction Growth and How to Overcome It—Friction in High Vacuum: Surface Cleaning by Rubbing—Friction in High Vacuum and at Very Low Temperatures—Friction at Very High Speeds: Frictional Heating—Polishing—Machining—Skiing on Ice and Metals—Friction at Very High Temperatures

VIII LUBRICATION   109
Hydrodynamic Lubrication—Aerodynamic Lubrication—Elasto-hydrodynamic Lubrication—Boundary Lubrication—The Breakdown of Boundary Lubricating Films—The Action of Boundary Lubricants—Greases—Extreme Pressure Lubrication

IX ROLLING FRICTION: BALL BEARINGS; AUTOMOBILE
TIRES; BRAKES; WOOD PULPING   131
The First Rolling Bearings—The Source of Rolling Friction—The Friction of Automobile Tires—A Few Words on Brakes—Pulping of Wood

Contents

X WEAR   147
The Non-existent Laws of Wear—Adhesive Wear—Mild Wear and Severe Wear—Effect of Environment—Effect of Speed—Surface Fatigue—Abrasive Wear—The Role of Dirt in Wear—The Complex Nature of Wear

XI SUMMARY AND PROSPECT: THE TRIBOLOGICAL CHALLENGE   157
The Past—The Present—The Tribological Challenge Confronting Industry—The Economic Challenge—The Tribological Challenge Confronting the Scientist: Materials—Lubricants and Lubrication—Dirt and Seals—The Prospect

INDEX   169

FRICTION

1 FRICTION IN EVERYDAY LIFE

> What is spherical moves easily; those bodies which are least stable and have the smallest surface of contact are furthered in their motion. Generally it is easier to further the motion of a moving body than to move a body at rest.
> Themistius (390–320 B.C.),
> *Physica*

What Is Friction?

It is a matter of common experience that if we place a book on a table it stays put. If we wish to move it we can do so in two ways: we can either tilt the table until the book begins to slide, or we can push the book until at a certain stage the push is strong enough to shift the book. The force resisting movement is due to friction between the book and the table and the force we apply to make it move is called the frictional force or more briefly the friction. The friction between bodies depends on several factors. For example, it is much easier to slide a book on a table than to slide a heavy chest over the floor. On the other hand, it is easier to slide a block of ice than a piece of wood of the same weight. Clearly friction depends on the nature of the materials as well as on their weight. Again it is much easier to move a chest over the floor if we support it on a couple of dowels. This is because, as Themistius observed 2,300 years ago, rolling friction is much smaller than sliding friction. It is this fact that made the wheel such an advance in the art of transport.

There are some situations where we need a quantitative knowledge of the friction. For example, in designing automobile brakes we need to know how effectively the brake will slow down the car: this is determined by the friction between the brake material and the brake drum. Again on braking on a wet or greasy road we need to know the friction between the tire and the road, for if we brake the wheels too effectively and lock them, the car may skid uncontrollably over the road. There is another automobile application where the friction may play a vital part—the door lock. This is an ingenious and intricate mechanism that has many moving parts and links, each of which introduces some frictional resistance. The lock tongue must hold firmly in the lock catch plate: yet the whole mechanism must be so free to move that it can be worked by a lady's finger. If the friction between the moving parts is too high, a person of normal strength will be unable to use the lock.

Two other examples show when we need to know how large the friction is. In a gyroscope a small cylindrical body is set spinning at a very high speed. To function satisfactorily the gyroscope must maintain a high and constant speed of rotation. For this reason friction must be as low as possible and extremely constant. At the other end of the scale there are the enormous turbines and generators, used in electric or hydroelectric power stations. The modern machinery used is well designed but there is bound to be some friction in the bearings. Although this may be relatively small the machines are so vast that the actual power lost in overcoming friction may be considerable. In that case a small reduction in friction may lead to an appreciable saving in power loss.

The Nuisance of Friction

In running machinery friction is undesirable and all sorts of attempts have been made to reduce it. One way is by using special low-friction materials, or by lubricating the surfaces with oils or greases. Another way is by clever design. For example, if we can convert sliding into rolling, the friction will be less. For this reason it is much easier to pull a cart or wagon that is mounted on wheels than to drag the cart if mounted on sledges.

In a modern automobile 20 per cent of the power is wasted in overcoming friction; in an airplane piston engine, 10 per cent; and in a modern turbojet 1.5 to 2 per cent is wasted. In these cases the energy loss is a nuisance but it is not the major trouble. The trouble is the damage done by friction, the wear or failure of some vital part. Apart from the wear itself, the friction can cause excessive heating. If we rub our hands together, for example, we use only part of the friction between our hands to rub off bits of skin: the main part of the work done in overcoming friction appears as heat. If we rub two pieces of wood together we can set them alight, as campers know. If we could look inside an aircraft brake when the plane is landing we would see that the brake and the brake drum get red hot. Even if the wear is slight these high temperatures will cause deterioration of the brake material and will tend to soften the metal of which the brake drum is composed. In the sliding of the piston rings inside the cylinders of an automobile engine high temperatures can cause failure of the lubricant and this in turn will lead to increased wear of the metal parts. The wear and heating produced by friction are probably the most important factors in limiting the capacity for work or shortening the life of aeroengines and other complex machines.

The Desirability of Friction

Although friction is generally considered a nuisance there are circumstances where the credit side is considerable, as anyone who has trodden on a banana skin knows. There are some situations where high friction is desirable, for example, in an automobile brake. Some other examples are:

(1) The interaction between shoe and floor. If there were no friction at all we could not walk with ordinary shoes or feet: we would have to use suction pads like an octopus to cling, step by step, to the floor.

(2) The interaction between an automobile tire and the road. Without friction we could not drive the car by making the wheels go round, they would simply skid. We would have to

use cogged wheels running over a corrugated rail like the rack and pinion used in funicular railways.

(3) The friction between the yarns in woven fabrics. The yarns are not held together by any glue but by the friction between them.

(4) The effectiveness of a knot as in knitting or as when two pieces of string or cord are tied together. It is the friction between the interlocking parts of the knot which hold it together. With a properly designed knot the harder we pull on the cord the higher the friction in the knot and the more it resists slipping. If the friction were too low the knot would not be able to grip. (Try tying two dead eels together.)

(5) The interaction between a nail and the wood into which it is hammered. The nail forces the wood apart so that it can enter. Consequently the wood presses on to the surface of the nail. If the friction were very low the nail would be squeezed out of the wood in the same way that we can squeeze a pip between our fingers and make it fly across the room.

(6) The grip between a nut and a bolt. When the nut is screwed up tight it presses against the sides of the thread of the bolt. It is the friction between the tightly squeezed faces of the threads that gives grip to the nut. Without friction the nut could not grip.

How We Define Friction: The Laws of Friction

So far we have been describing friction in a qualitative way. How can we describe it more quantitatively?

Suppose an engineer has to make a piece of machinery in which one piece of metal has to slide over another. There may be several materials that he can choose from and he wants the combination that will give the smallest friction. He could build the whole piece of machinery with all the different combinations available and compare them, a costly and time-consuming business. A much simpler method is to carry out a series of tests as shown in Figure 1. We place a block of metal A on top of surface B; attach a string to A, pass it over a pulley and fix a scale pan at the lower end. We can then add weights to the pan until body A begins to move. The

Friction in Everyday Life

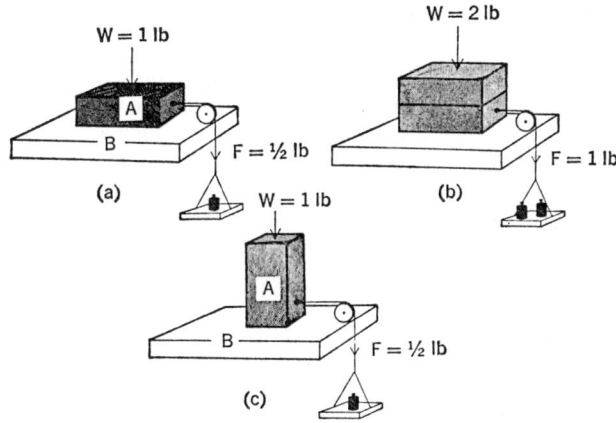

Figure 1 *A simple method of measuring the friction between body A resting on surface B. By adding weights to the scale pan we are able to apply a force F sufficient to start A sliding on B. This is the static friction (or static frictional force) between A and B. (a) When the weight W of body A is 1 lb. we find that a weight of ½ lb. in the pan is sufficient to start A sliding. (b) If the weight of W is doubled, the force required to produce sliding is doubled. This illustrates the first law of friction: frictional force is directly proportional to the weight of the body. (c) If body A is turned to stand on its small face the friction is the same as in (a). This illustrates the second law of friction: friction does not depend on the area where the two surfaces are in contact.*

weights in the scale pan plus the weight of the scale pan itself is then the force F overcoming friction.

Suppose body A weighs one pound and we find that the force to overcome friction is half a pound (Figure 1a). We could then say that the frictional force F is half the normal load W pressing body A onto B. If we were to put a similar block on top of A so that the two weighed 2 pounds (Figure 1b) we would find that the force to slide the pair over B would be roughly 1 pound, that is, half the normal load. If we increased the load on A to 5 pounds we would find that F would have to be about 2½ pounds to produce sliding, that is, it is again one half the normal load. This illustrates what is called the first law of friction: frictional force is proportional to the normal load.

We should also find, and this has always surprised engineers, that the frictional force will not depend on which way block A is rested on surface B, whether with its small face or its large face in contact (Figure 1c). This is known as the second law of friction: friction does not depend on the apparent area of the contacting solids, that is, it is independent of the size of the bodies. We can generalize the results of all the experiments carried out on bodies A and B by saying that "the frictional force of A on B is half the normal load" or

$$F = \tfrac{1}{2}W.$$

The next step is simple and shortens our way of describing the friction. We take the *ratio* of the friction to the load. In the case described above we get

$$F/W = \tfrac{1}{2}.$$

The ratio is defined as the coefficient of friction and is given the Greek symbol μ (mu). We write

$$\mu = \text{Ratio of frictional force to normal load} = F/W.$$

In the above case μ is $\tfrac{1}{2}$. The engineer then knows that for this combination of materials if the normal load between the bodies is W pounds the force to move it will be $\tfrac{1}{2}$W pounds. If he then finds a combination of materials for which $\mu = \tfrac{1}{3}$, the frictional force will be $\tfrac{1}{3}$W; and so on. In general the force to move the body will simply be

$$F = \mu W.$$

As Themistius pointed out over 2,000 years ago, it is harder to start a body moving than to keep it moving. The force to start the body moving is called the static friction (F_s) and the force to keep it moving, the kinetic friction (F_k). There will be two corresponding coefficients of friction μ_s and μ_k where generally μ_s will be greater than μ_k but we shall not pursue this further here.

Friction in Everyday Life

Some representative values of coefficients of friction for bodies sliding on one another (μ_s) are:

Book on table	$\mu = 0.3$
Brake material on brake drum	$\mu = 1.2$
Dry tire on dry road	$\mu = 1$
Wet tire on wet road	$\mu = 0.2$
Copper on steel, dry	$\mu = 0.7$
Ice on wood	$\mu = 0.05$

In rolling and rotary motion there is also a coefficient of friction but this has to be defined slightly differently (see Figure 2). If the weight of the wheel or roller is W and we have

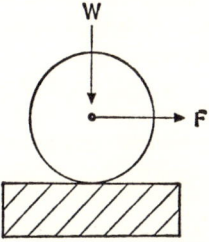

Figure 2 *A roller or cylinder of weight W rests on a flat surface. A force F applied at the center is just able to make the cylinder start rolling. F is then the rolling friction force and the ratio F/W is the coefficient of rolling friction.*

to apply a force F at the center to enable rolling to take place we can define the coefficient of rolling friction as $\mu_R = F/W$. For a hard steel roller on a hard steel surface μ_R may be as little as 0.001, that is, the rolling force is only one thousandth of the normal load. This situation operates in ball and roller bearings. With more deformable materials the rolling friction is higher but even for a rubber tire on a road the rolling friction is 0.05. Clearly it is advantageous to convert sliding motion into rolling motion.

What Causes Friction? A New Word: Tribology

What is the cause of friction and what is really happening

at the interface between solids during sliding? This problem exercised the intellect of the ancient philosophers, the genius of Leonardo da Vinci, the solid military mind of Coulomb and the analytical and practical skills of the nineteenth-century engineers. But it is only in the last thirty or forty years that physicists, chemists, metallurgists and engineers have tackled the problem in a more effective way. The study of friction demands an interdisciplinary approach because friction is the result of a number of interacting processes. Although friction is simple to measure it is complicated to explain. In recent years as a result of increased interest in friction, lubrication and wear, a new word has been coined to describe the field: tribology. This word is derived from the Greek *tribos,* which means "rubbing." Tribology is defined as "the science and technology of interacting surfaces in relative motion and of the practices relating thereto."

Friction arises from the interaction of solids at the regions where they are in real contact. To understand it we need to study the shape and contour of surfaces, the way they deform when they are pressed together, how solids adhere, the strength properties of the interface, the role of surface films, and how energy is lost when the surfaces are deformed during sliding.

Before dealing with the nature of solid surfaces and the way in which they interact when placed in contact and slid together we shall first discuss some of the more important ideas about friction that have appeared in the earlier scientific literature.

II FRICTION IN HISTORY.
EARLY IDEAS—GOOD AND BAD

> Whether the friction of the heavens makes a sound or no: . . . if the heavens are not smooth at the contact of their friction it follows that they are full of lumps and rough, and therefore their contact is not continuous and if this is the case the vacuum (between the lumps) is produced which it has been concluded does not exist in nature. We arrive therefore at the conclusion that the friction would have rubbed away the boundaries of each heaven . . . and then there would not be friction any more and the sound would cease and the dancers would stop . . .
> Leonardo da Vinci (1452–1519),
> *The Notebooks,* F 56 V

The First Ideas

Ever since man tried to drag loads over the ground he has been aware of the existence of friction. He may not have known how to explain it but he knew how to reduce it. For example, a mural painting in a grotto at El Bersheh (c. 1900 B.C.) shows a colossus being pulled along on a sledge while one man stands in front of the sledge and pours a lubricating oil in its path (see Figure 3). Again, a chariot of about 1400 B.C. was found in an Egyptian tomb with remains of the original lubricant on its axle. As one might expect, the classical philosophers such as Aristotle left comments showing that the existence of friction, the usefulness of lubricants, the advan-

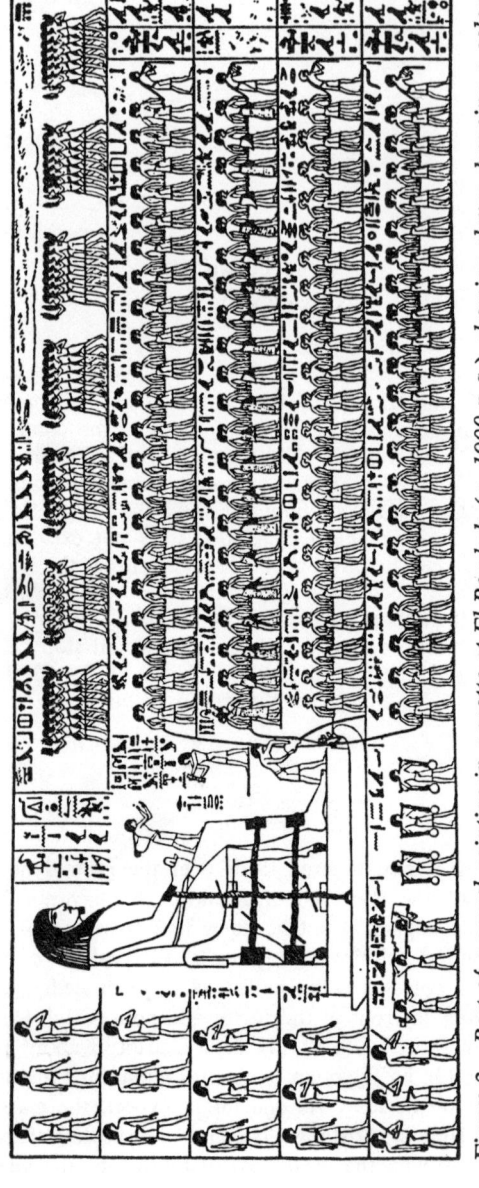

Figure 3 Part of a mural painting in a grotto at El Bersheh (c. 1900 B.C.) showing slaves dragging a colossus on a sledge while one man pours lubricating oil in its path.

Friction in History

tages of metal facings in bearings were common knowledge among the ancients. But nothing of a scientific nature was undertaken for almost 2,000 years.

In the middle of the fifteenth century Leonardo da Vinci deduced the two basic laws of friction two hundred years before Newton had given a clear definition of force. "Friction," he wrote, "produces double the amount of effort if the weight be doubled," that is, the friction is proportional to the load. This is the first law of friction. He also observed that the area of the surfaces in contact had little effect on the friction; this is the second law. A typical page from his notebooks is shown in Figure 4. It is seen that his friction experiments are very similar to those described in Chapter I. His friction measure-

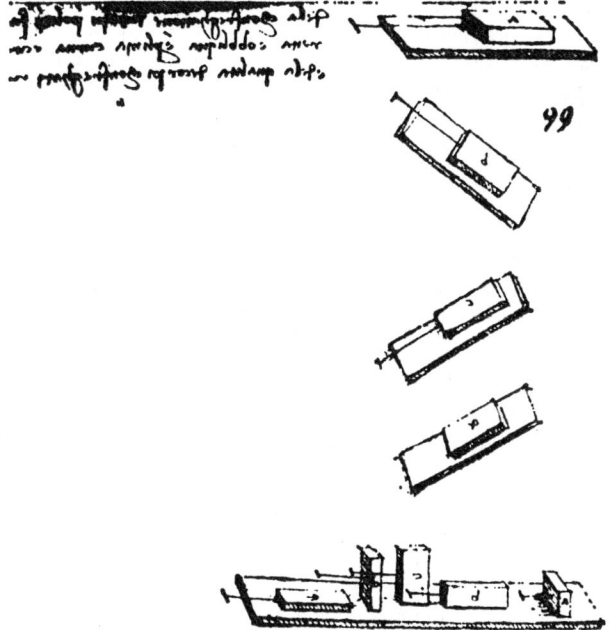

Figure 4 *A page from the notebooks of Leonardo da Vinci showing the experimental arrangement by which he showed that friction is independent of the size of the surfaces making contact.*

ments showed in a quantitative way what had long been known through practical experience, that various materials "move with different ease." This, he said, was because their friction is different; in his view bodies with smoother surfaces have smaller friction. He even produced a quantitative result, that "every frictional body has a resistance of friction equal to one quarter of its weight"; that is (in modern terms), the coefficient of friction is one quarter. For surfaces which are not too clean this is a fair observation.

Leonardo da Vinci was one of those exceptional figures escaping easy classification and simplified theories of human personality. Much of his scientific work was stimulated by the problems and interests surrounding him; much—including friction—arose mainly from the scientific curiosity of an outstanding perceptive intellect. His work on friction and its conclusions were forgotten, and it was not until 200 years later that interest again arose in this area. During these two centuries science had taken on a "new look": it had become modern. Newton had conquered the simple laws of forces, reactions, accelerations and momenta. The great era of classical mechanics had begun.

Academies of Science and the Industrial Revolution

There were two developments in the seventeenth and eighteenth centuries which had a profound effect on the progress of science and technology; probably they were interrelated. One was the establishment of Academies of Science in most of the larger European countries and America. The Royal Society of London was founded in 1662, the French Academy of Science in 1666, the Prussian Academy in 1700, the Danish in 1742, the St. Petersburg Royal Academy in 1725, the American Philosophical Society (founded by Benjamin Franklin) in 1742. The Royal Houses of Europe cultivated and encouraged the sciences as a matter of prestige and pride. Members of royalty, of the aristocracy and of the educated classes in general often took a direct personal interest in scientific affairs. A man of culture would include science among his accomplishments.

The second factor was the growth of mechanically driven

Friction in History

machinery and the beginning of the industrial revolution. Until that time primary power had been provided by men, horses, windmills and crude water mills. Now engineers began to look for better forms of motive power, especially water wheels and primitive steam engines. The water wheel was greatly improved. Denis Papin (1647–1712), a French scientist who later made his home in Germany and in England, constructed the first steam engine which used a *piston in a cylinder*. In 1705 the English engineer Thomas Newcomen developed the first practical steam engine, which was widely used for the draining of mines and for the pumping of water for supply purposes. In 1770, Captain Cugnot, a French artillery officer, built the first self-propelled road vehicle—a massive steam-driven affair with a top speed of 3 m.p.h. It was in this atmosphere that the first modern research in friction began.

Amontons and the Early French Workers

The first scientific studies of friction were carried out in France. One factor may have been military needs; another, the popular use among the leisured classes of ornamental fountains for which pumps were required. For example, the engineer M. Parent (1666–1716) published a two-volume work on hydraulics and included in it a critical and thoughtful review of friction as it was understood in his day. However, the first original work on friction was due to Guillaume Amontons (1663–1705). By training he was an architect but in those days specialization in one area did not exclude interest in the whole field of science. Amontons developed the first gas thermometer using the expansion of air as a means of measuring temperature. His work in this field led him to design (but not to construct) a steam engine, and he remarked rather charmingly that it ought to be an economic success, for unlike horsepower the engine would not need feeding when it was not in use. He developed other scientific instruments and in 1699 published a paper in the Proceedings of the French Royal Academy of Sciences on friction. In this work he rediscovered the two forgotten laws of friction originally derived by Leonardo da Vinci. The first law, that the frictional force is proportional to the normal load was accepted by the Academy without question. The second law, that friction is independent of the size of the bodies, was viewed by the Academy with astonishment and skepticism. They instructed their senior academician De la Hire (1640–1718) to repeat Amontons' experiments and check their validity. This he did and confirmed Amontons' conclusions. Amontons' laws of friction have remained with us to this day as a very good working approximation.

Amontons recognized that the surfaces he worked with were not smooth; on the contrary, they were rough even to the naked eye. He thought friction arose from the work done in lifting one surface over the roughnesses on the other, or from bending the roughnesses down, or from breaking off the roughnesses. In the century that followed, most scientists accepted the view that friction in one way or another was due to the roughnesses on the surfaces.

Desaguliers' Ideas of Adhesion in Friction

A very different view of the mechanism of friction was put forward thirty-five years after Amontons' paper by an English scientist John Theophilus Desaguliers (1683–1744). Desaguliers was the son of a French Protestant priest who escaped from Paris (after the revocation of the Edict of Nantes in 1685), reputedly by hiding in a barrel. The son took his degree at Oxford and taught physics there for several years. In London he was friendly with Isaac Newton. Later he became chaplain to the Prince of Wales, for it was an age when educated men felt equally at home in all branches of learning. In 1734 he published *A Course of Experimental Philosophy*, which contains many quaint as well as precise scientific observations. For example, he estimates that the power of a horse is equivalent to that of five Englishmen or seven Frenchmen or seven Dutchmen. However, he redeems himself from racial pride by remarking that Turkish porters can carry double the load of an English porter. Again, he points out that, for optimum strength, spoked wheels should be dish-shaped so that if the wheel falls into a rut the spokes taking the compressive stress are more nearly vertical. In discussing friction Desaguliers considers the surface roughness theory and points out that

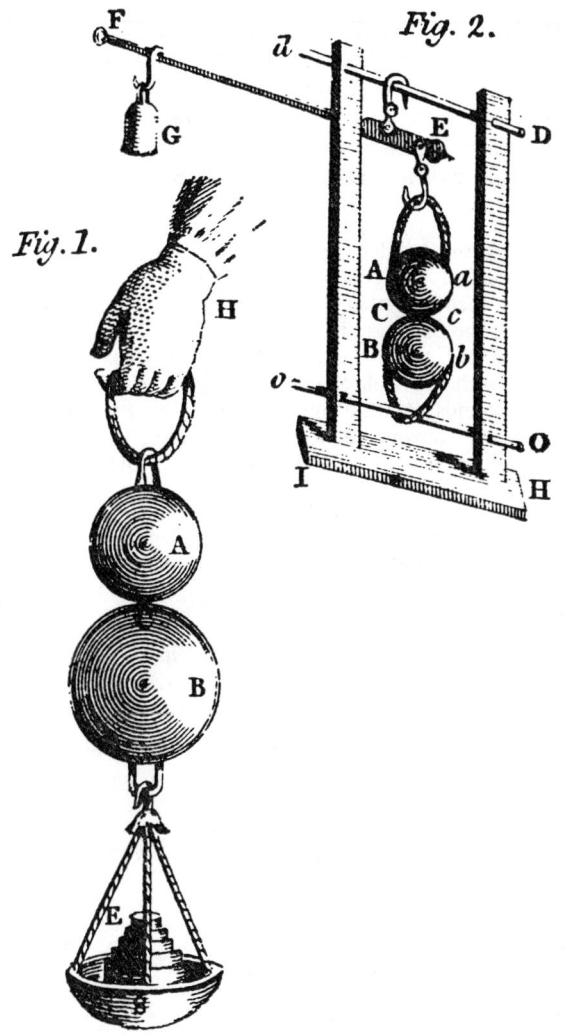

Figure 5 *Figures from a paper published by Desaguliers (1683–1744) on the adhesion of lead.*

as the surfaces are made smoother they ought to slide more easily, "yet it is found by Experience that the flat Surfaces of Metals or other bodies may be so far polished as to increase Friction." He attributed this paradoxical behavior to the adhesion between the surfaces at the regions of contact though in his terminology it is called cohesion. He then goes on to describe his earlier work on the adhesion of lead (see Figure 5).

> I took the Leaden Balls . . . and having from each of them cut off a Segment of about ¼ inch in Diameter, I pressed them together with my Hand, with a little Twist, to bring the flat parts to touch as well as I could. The Balls stuck so fast, that . . . the lower one was sustained though loaded with the Scale S and the Weights E which amounted to 16 pounds. A little more Weight added separated them and, upon viewing the touching surfaces, it appeared that they did not exceed a Circle of $\frac{1}{10}$ Inch Diameter. . . .

In some cases the adhesion was as great as forty-seven pounds. Although Desaguliers provided a different approach to friction his ideas were not fully developed. He recognized that adhesion played a part in friction but could not see how he could explain the laws of friction.

A similar confusion evidently existed in the mind of Coulomb, for he considered the possibility of adhesion but rejected it because adhesion would imply that the friction would double if one doubled the number of contact regions, that is, if one doubled the area of contact.

Coulomb's Classical Study: The Importance of Surface Roughness

Charles Augustin Coulomb (1736–1806) was by training an engineer. He was a captain in the French Royal Corps of Engineers and served for nine years in the West Indies (Martinique), where he helped to build Fort Bourbon. He returned to France because of illness in 1779 and began his research on friction. This was stimulated by a prize offered by the French Academy of Sciences for research that would be useful in the design of machinery. The Academy stipulated

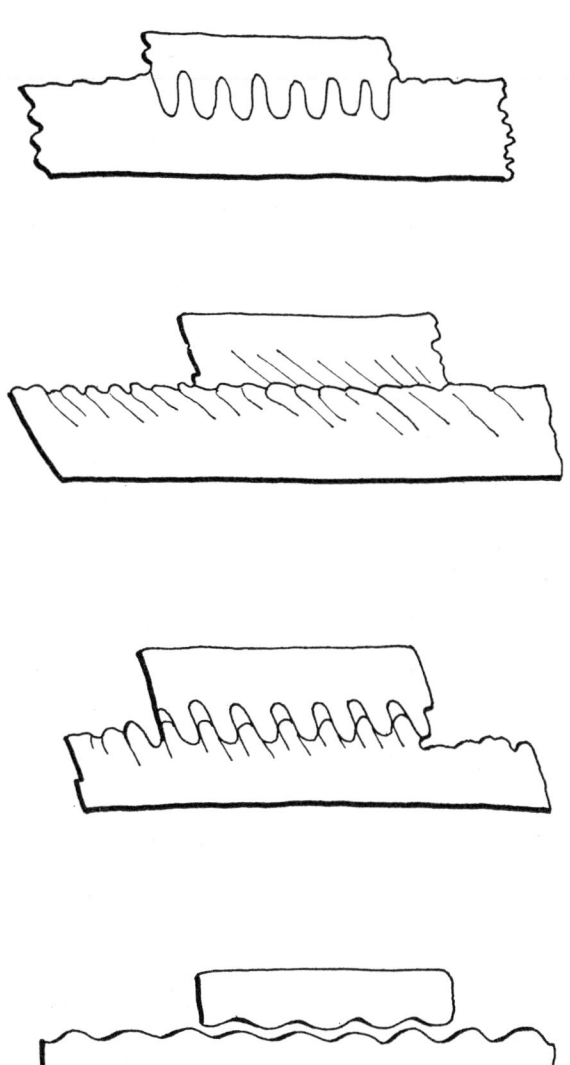

Figure 6 *Part of an illustration from Coulomb's book* Theory of Simple Machines, *published in 1781, which shows his ideas on the interlocking of surface roughnesses.*

that "the laws of friction and the effect of stiffness in cords should be determined by new experiments on a large scale: and demanded further that these experiments should be applicable to machines used in ships such as the pulley, the capstan and the inclined plane." The last item refers to slipways in the launching of ships. In 1781 Coulomb published his famous *Theory of Simple Machines,* which won him the Academy award; he was also elected to the Academy in that year. During 1784–89 he wrote his memoirs on electricity and magnetism and established the law of force between electrical charges. Science acknowledges his contribution in this field by naming the unit of electrical charge after him—the coulomb. On the outbreak of the Revolution he gave up his public office in Paris and retired to the country but, around 1795, in the days of the Consulate he was appointed Inspector General of the University of Paris.

Today Coulomb's work on friction seems dull, but his conclusions are clear and forthright. Coulomb, like Amontons and his contemporaries, recognized that most of the surfaces they worked with were not smooth. When two such surfaces are placed together the contact resembles the interlocking bristles of two brushes placed against one another. Perhaps a better analogy, as can be seen in Figure 6, from Coulomb's paper, is that the roughnesses fit together like two pieces of a jigsaw puzzle. Thus the area of contact would be larger for larger surfaces. If friction was due to adhesion at the interface it would be larger for larger bodies, that is, the friction would depend on the size of the bodies. Coulomb's own measurements showed that this was not so. Although he recognized that adhesion might play some part in friction, he rejected it as the main cause. He felt friction was due to the role of surface roughnesses, to the work done in dragging one surface up the roughnesses of the other.

Suppose the lower surface consists of a single roughness which makes an angle θ to the horizontal (Figure 7). Suppose the upper surface resting on it carried a load W acting downwards. What is the horizontal force F that would be required to pull the upper surface up the slope? The simplest way of working this out is in terms of work. Two principles are involved. The first is that if a force X moves a distance x

Friction in History

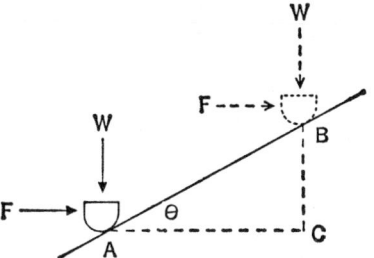

Figure 7 *A body of weight W rests on a surface making an angle θ to the surface. The horizontal force F necessary to push the body up the slope against gravity moves a horizontal distance AC, while the load W moves a vertical distance BC. For reasons discussed in the text F.AC = W.BC, so that F is simply equal to W tan θ.*

in the direction of the force the work done by the force is the product Xx. Now consider what happens if the top body is pushed from A to B. The horizontal force F moves a horizontal distance AC, so that the work done is F.AC. The load W is moved a vertical distance BC; the work done on the load in overcoming gravity is W.BC. We now come to the second principle of work. In any system where energy is not lost the work done in changing from state 1 to state 2 does not depend on the path followed but only on the initial and final state. Consequently in moving the body from A to B

$$F.AC = W.BC$$

$$F = W\frac{BC}{AC} = W \text{ tangent } \theta$$

so that if θ is constant F is proportional to the load W.

Hence $F/W =$ tangent θ.

It will be remembered from Chapter I that F/W is the formula for finding the coefficient of friction μ. Therefore

$$\mu = \tan \theta.$$

We see, first, that in this model the coefficient of friction does not depend on the load. Hence the frictional force F is proportional to the load W (first law). Secondly, the size of the bodies does not come into the picture. If the weight of the top body is W and the average slope of the roughnesses is θ the coefficient of friction is determined only by tan θ. The size of the bodies is irrelevant (second law). These two features were the major attractions of the "roughness theory" of sliding friction. To some extent they receive support from our intuitive feeling that rough surfaces must have a greater friction than smooth ones. Intuition is not enough; as the quotation from Desaguliers' paper shows, smooth surfaces can have a higher friction than rough ones. In fact, surface roughness is no criterion as to whether the friction will be higher or lower.

There is one point in Coulomb's theory which we have glossed over. What should we use for θ if the slopes of the individual roughnesses vary over a big range? Should we take the steepest slope on the grounds that it will gradually lift the upper surface away from the gentler slopes? Probably the average slope is as good an assumption as any. What then of those slopes which face downwards as well as upwards?

How Is Energy Dissipated?

How, on Coulomb's theory, is energy used up in overcoming friction? Let us consider a single region of contact where the slope is θ and the vertical load is W_1 (Figure 8a). If we try to slide this up the surface there is a component of gravity down the slope resisting movement. If there is no adhesion this is simply the force F_1 required to overcome gravity. The coefficient of friction is $\mu = \tan \theta$. If now the whole body makes contact at a number of regions and they are all being dragged up identical slopes (Figure 8b) the overall coefficient of friction for the whole body is again tan θ. But what happens a little later when the contact regions have moved to the other side of the slope? Clearly the body slides along on its own and indeed itself *exerts* a horizontal force in the direction of sliding (Figure 9). A little thought shows that the work that we have to use in sliding up the slopes is restored to us in sliding down the slopes. This is an

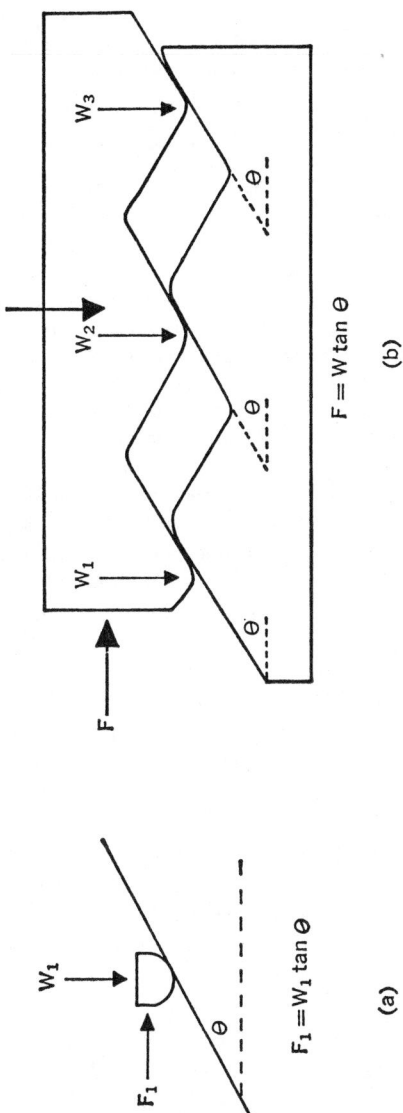

Figure 8 (a) A body of weight W_1 rests on a surface inclined at an angle θ; the horizontal force F_1 required to push the body up the slope against gravity is $F_1 = W_1 \tan \theta$. (b) A body making contact at a number of regions exerts vertical forces W_1, W_2, W_3, etc. at these regions, the total weight W of the body being $W = W_1 + W_2 + W_3$. If the slopes at each contact region are the same and equal to θ the total horizontal force F to produce sliding will be $F = W_1 \tan \theta + W_2 \tan \theta + W_3 \tan \theta = W \tan \theta$.

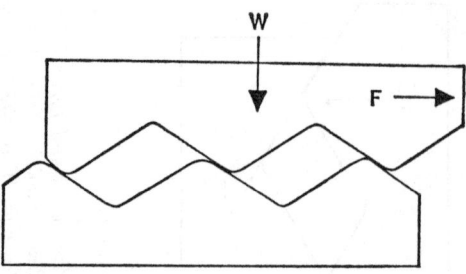

Figure 9 *The situation of the body shown in Figure 8b after it has moved a little so that the contact regions are on the other side of the slopes. Gravity now acts down the slopes in the direction of sliding so that the body is urged* forward *by gravity. The work done in pushing the body up the slopes is recovered by the body when it slides down the slopes.*

extension of the principle of work described earlier. If friction is due only to the overcoming of gravity, we can only do a net amount of work if we raise one body by a definite distance. But if the average position of the top body is in a horizontal straight line there is no net vertical movement. Parts of the body may be sliding up roughnesses but other parts will be sliding down equivalent slopes somewhere else or at a later instant. The frictional force would therefore fluctuate between positive and negative and its average value would be zero. Consequently no energy could be lost in moving one body over another in a horizontal plane. This criticism of the roughness theory was put forward in 1804 by John Leslie (1766–1832), Professor of Physics in the University of Edinburgh, and remains unanswered. Of course one can say that work is used up in dragging a body up a slope and when it gets to the top edge of the slope it falls with a bang, bending and denting the surfaces, so that all the work done on it is lost as deformation work during impact. If we adopt this view, however, we have gone a long way from the dragging-up-the-roughness model. We are really talking of a deformation mechanism. Indeed Leslie himself believed that friction was due to the work done in deforming the surface of one body by the roughnesses on the other, a mechanism which, in some circumstances, is still considered to play an important part in the sliding friction of solids.

Leslie opposed the view that friction could arise from surface adhesion: "a perpendicular force acting on a solid can evidently have no effect to impede its progress." Leslie's view here is that a vertical force (adhesion) can have no horizontal (frictional) component. This is correct if judged in terms of forces and components, but Leslie's error arises from not asking the next question. If his statement were true, a glued joint would have no shear strength. Certainly the lead spheres in Desaguliers' experiments not only stuck together with a force which, in some cases, was so large that a normal force of forty-seven pounds was required to pull them apart: a very large tangential force would have been required to shear the region where the spheres had welded or adhered together. In other words, the adhesion of Desaguliers' lead spheres involves a tangential resistance to sliding. In fact, adhesion can play a major part in the sliding friction of surfaces particularly if they are fairly clean.

Surface Roughnesses and the Area of Contact

There is another major defect in the early work on friction: it is concerned with the area of contact. The great contribution of the French School was to emphasize that contact only occurs at discrete points, where the roughnesses or asperities on one surface touch the other. But they regarded this as a sort of geometric matching of the roughnesses in the way that two pieces of a jigsaw puzzle fit together. Thus the area of contact depends on the overall size of the bodies and is quite independent of the load. It was because of this that Coulomb rejected the adhesion mechanism of friction. Nearly 150 years passed from the time of Coulomb before it was realized that the individual asperities are deformed according to laws of elastic and plastic deformation which were unknown in his day. It was this relatively recent work that showed that the true area of contact depends both on the geometry of the surfaces and on the way in which the individual asperities are deformed. And it was this approach which made it possible to give adhesion its rightful place in explaining the friction of sliding surfaces.

It is important to understand what a solid surface looks like and how it is deformed when a similar surface is pressed on to it and then slid over it. This we shall do in the following two chapters.

III SOME PROPERTIES OF SOLIDS —AND OF LIQUIDS

All these things consider'd, it seems probable to me, that God in the Beginning form'd Matter in solid, massy, hard impenetrable, movable Particles. . . .
Now the smallest Particles of Matter may cohere by the strongest Attractions, and compose bigger Particles of weaker Virtue . . . and so on for divers Successions, until the Progression end in the biggest Particles . . . which by cohering compose Bodies of a sensible Magnitude. If the Body is compact, and bends or yields inwards by Pression without any sliding of its Parts, it is hard and elastick, returning to its Figure with a Force rising from the mutual Attraction of its Parts. If the Parts slide on one another, the Body is malleable or soft. If they slip easily, and are of a fit Size to be agitated by Heat, and the Heat is big enough to keep them in Agitation, the Body is fluid. . . .

Isaac Newton (1642–1727),
Opticks, Book Three, Part I

Why Is a Solid a Solid?

All matter is made of atoms: these are the unit building-blocks. Atoms sometimes exist on their own; for example, there are single atoms of copper in a piece of copper; there are also single atoms in the neon gas that fills fluorescent lamps. More often atoms combine with themselves or with other atoms to form stable units known as molecules. For example, oxygen gas consists of molecules of oxygen, each oxygen molecule consisting of two oxygen atoms joined together by chemical bonds. Again, two atoms of hydrogen and one atom of oxygen join together to form one molecule of water. One atom of sodium joins up with one atom of chlorine to form a molecule of sodium chloride (common table salt).

All molecules and atoms attract one another. This is primarily the result of the electrical structure of the atom. There is, of course, also a gravitational force but this is billions of times weaker than the electrical interatomic forces. We do not need to go into this further except to say that it is the strong attractive force between atoms (or molecules) that gives a solid its solidity and strength. These forces pull the atoms and molecules together until they are very tightly packed. We may for simplicity regard the atoms or molecules as hard incompressible units, although in fact they must deform a little when they are pressed together: this produces a repulsive force which just balances the attractive force, for when the solid is in equilibrium there can be no net forces remaining between the atoms. For example, if we press two golf balls together we may think that there is a resultant force equal to the force with which we press them together. But in practice they will deform a little at the contact region until the forces resisting further deformation are just equal to the pressing force. There is indeed a repulsive force between the golf balls equal to the external force we apply. Of course this repulsive force only operates when the balls are in contact and this is true of the repulsive forces between atoms: they are called "short-range" forces. On the other hand, if we put a small powerful magnet into one of the golf balls and a piece of soft iron into the other golf ball they would attract one another over quite a long distance. This resembles the relatively "long-range" attractive forces between atoms (or molecules). If our two balls were now allowed to move freely they would come together under the attractive force, press on one another as a result of this, and then produce a slight deformation at the contact region until the repulsive force balanced the attractive force.

There are different types of attractive forces between atoms and molecules. There is the force which occurs in metals (the metallic bond), that which occurs in a crystal of table salt

Some Properties of Solids—and of Liquids

(the ionic bond) and that which occurs between molecules in a piece of plastic, like nylon or polythene (the van der Waals bond). If we wish to deform a solid, that is, to change its shape, we have to apply forces to compete with the forces between the atoms and molecules.

The Size of Atoms: How to Sketch the Forces Between Them

How big are atoms? The unit of length used in describing the size of an atom is the angstrom, named after A. J. Ångström and written shortly as Å: it is one hundred millionth of a centimeter (10^{-8} cm). For example, the diameter of a copper atom is about 3 Å, so that if we laid 100,000,000 copper atoms in a row they would just occupy three centimeters. Clearly atoms are very small indeed.*

We could describe the forces between two copper atoms by means of Figure 10, where we have plotted force against the distance between the atoms. The repulsive forces only operate when the atoms are just or almost in contact and they increase rapidly as we squeeze the atoms closer together. Consequently the curve A, describing this, is very steep. On the other hand, the attractive forces vary more gently with distance. They fall off with increasing separation, and for distances greater than 4 or 5 atomic diameters (say, 12 to 15 Å) they are relatively unimportant (curve B). The curve describing the resultant force between the two atoms (heavy line C) is the sum of these two forces. Similar curves are obtained for the force between two oxygen molecules or between a benzene molecule and a metal atom. Indeed the force between all atoms and molecules as a function of separation is of the same shape as that shown in Figure 10 although of course the scales on the horizontal and vertical axes will vary considerably.

The force-separation curve always contains two main fea-

* Even a molecule of rubber, which may contain 100,000 atoms of carbon in a continuous chain, would only be a thousandth of a centimeter long if we could stretch it straight: but it would only be 5 Å thick, so we should never be able to see it with an optical microscope.

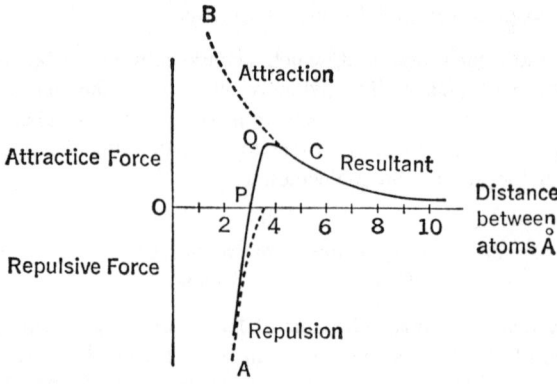

Figure 10 Sketch showing the way in which the force between two atoms varies with the distance between them. There is an attractive force which is weak for large separations (say, 10 Å) and becomes stronger as the atoms get closer together. There is also a repulsive force which only becomes important as the atoms are brought into contact (separation less than, say, 3 Å). The thick line is the resultant. The attractive and repulsive forces are in equilibrium when the resultant force between them is zero (point P). The distance OP is the equilibrium separation.

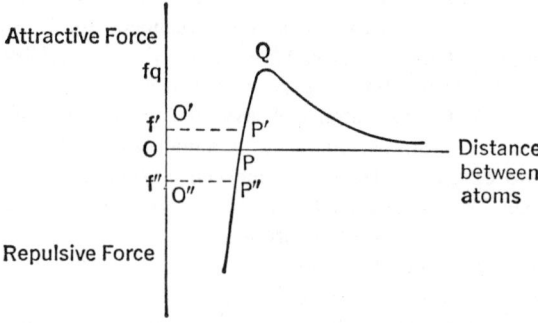

Figure 11 Force-separation curve for two atoms. OP is the equilibrium separation between them. We may increase the separation of the atoms from OP to O'P' by pulling with a force f'. Similarly we may reduce the separation from OP to O"P" by pushing the atoms together with a force f". If we pull with a force f_q we shall pull the two atoms apart at Q.

Some Properties of Solids—and of Liquids

tures and to bring these out more clearly we have redrawn the curve in Figure 11. The first concerns the point P, where the resultant force is zero, that is, the attractive and repulsive forces just balance. This is the equilibrium separation between the two atoms (or molecules) under consideration. The force-separation curve at this point is very steep. The second feature concerns the point Q, which we shall discuss when we deal with fracture.

Elastic Deformation

Consider two atoms for which the force-separation curve is as shown in Figure 11. Suppose we wish to pull the atoms further apart so that their separation is increased from OP to O'P'. We should have to pull with a force f' corresponding to the length OO' on the force axis. Similarly if we wish to push the atoms together to decrease the separation from OP to O''P'', we should have to push with a force f'' corresponding to the length OO'' on the force axis. If we were to remove the force the atoms would jump back to their old equilibrium separation OP. This sort of change in separation which is reversible when the external force is removed is known as elastic deformation.

Consider, for example, a strip of spring steel. If we clamp one end on a table (Figure 12) and hang a weight on the

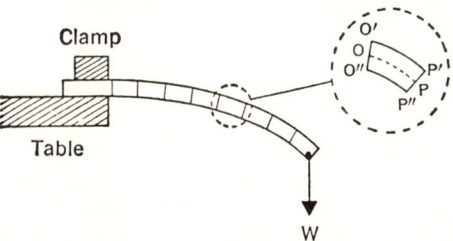

Figure 12 *Bending of a steel strip, clamped at one end and weighted at the other. Careful examination of any portion of the strip (see insert) will show that the central portion OP is unchanged in length, the outer portion stretched to O'P' and the inner portion compressed to length O''P''.*

free end we shall bend the strip as shown. If we examine any portion of the beam we will find that the central layer of the strip OP is unchanged in length while the outermost layers have been stretched to a length O'P', the inner layers compressed to a length O''P''. All the atoms in the outermost layers have been pulled further apart, all those in the inner layers have been pushed together. The atomic forces involved (derived from Figure 11) can be used to calculate exactly how much the strip will bend under a given weight W. If the strip is not bent too far it will spring back to its original shape when the weight is removed and the atoms will return to their equilibrium separation.

A further characteristic of elastic deformation is that we can recover the energy expended in deforming the solid. Consider, for example, the case discussed above of the spring steel strip. When we hang the weight on the end of the strip it moves vertically downward and performs a certain amount of work in deflecting the strip. If we now couple the spring to a suitable mechanism and remove the applied weight so that the strip returns to its original shape, we find that we get back the same amount of work (almost) as we expended in bending the strip. We have inserted the word "almost" because no real solid is 100 per cent elastic: a certain amount of energy is always lost even in elastic deformation. With hard steels the amount lost is extremely small. In a watch spring, for example, we can wind the spring thousands of times and it will unwind and drive the watch, furnishing the watch mechanism with (almost) all the energy we expended in winding it up.

Plastic Deformation

If the stretching or compression of the atoms in a solid does not change their separation by more than, say, 1 per cent, the deformation is elastic, that is, it is reversible. If, however, (still considering the behavior of a strip of spring steel), we bend the strip too far, so that we demand too great an extension or compression of the atomic spacing, we shall find that we have exceeded the "elastic limit" of the material. The strip does not return to its original shape when the weight is removed: it will have a permanent bend. What happens here

Some Properties of Solids—and of Liquids

is illustrated in Figure 13. Consider two rows of atoms in a part of the metal where the rows are in perfect register. This occurs, for example, in a crystalline part of the metal (see below). If we pull the top layer over the bottom layer (Figure 13b), we increase the distance between the nearest atoms and so have to compete with the interatomic forces. If the amount of displacement is small the atoms can slip back when the force is removed. If, however, we pull too hard we shall find that one row will slip over the other and when the force is

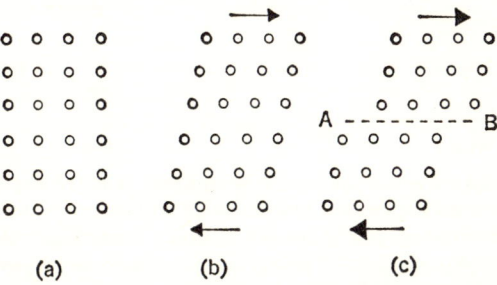

Figure 13 *Effect of applying a tangential (or shear) force on a crystal. (a) Initial arrangement of atoms. (b) Arrangement when a moderate shear force is applied. The crystal is distorted but if the applied force is not too large the crystal will return to its original shape when the force is removed. The deformation is elastic. (c) Arrangement if the shear force is too large. There is slip along AB between the neighboring sheets of atoms and when the force is removed the crystal remains permanently deformed. The deformation is plastic.*

removed one row will be left displaced relative to the other (Figure 13c). The material is permanently deformed although it is still a single piece of material. A similar process occurs in tension as is shown schematically in Figures 14a and 14b. This type of deformation, which always involves the slip of one plane of atoms over another, is known as plastic deformation and occurs whenever we try to increase the separation between atoms by more than 1 per cent or so. Some materials such as gold and silver can undergo very large amounts of

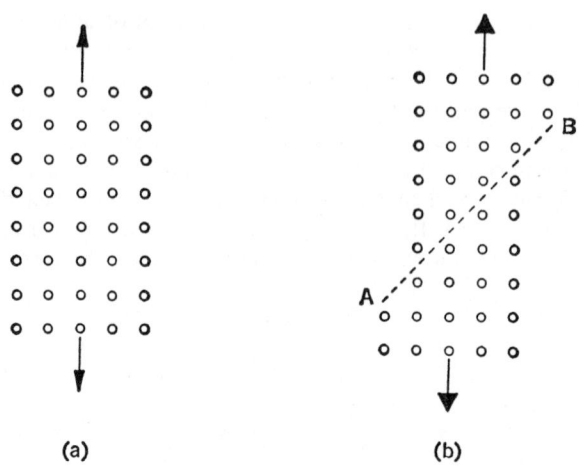

Figure 14 *Crystal subjected to tension.* (a) *If the tensile force is small the crystal returns to its original shape when the force is removed.* (b) *If the tensile force is too large, slip will occur along some appropriate direction and the crystal will remain deformed when the force is removed.*

plastic deformation before they finally fracture (see below). This ability to flow plastically is known as "ductility."

There are two characteristics of plastic deformation that distinguish it from elastic deformation. First, the energy expended in producing the deformation cannot be recovered when the deforming force is removed. The energy is lost and the major part of it appears as heat. If, for example, we bend our strip of metal forward and backward several times, plastically, we can easily feel the temperature rise around the part of the bend where plastic deformation has occurred.

Secondly, for reasons that we cannot go into here, the process of plastic deformation generally makes the material more resistive to further plastic deformation. This is known as "work-hardening." As the material work-hardens it becomes less and less ductile.

Fracture

If we bend our strip even further it may snap into two pieces: we have fracture. We would find the same behavior if we suspended our strip of metal vertically with its top end fixed and hung weights at the bottom end. For small weights the strip would extend elastically (up to 1 per cent extension) and on removing the weights it would return to its original length. If we increased the weight the strip would extend plastically by the slipping of atoms over one another: the extension would remain on removing the weight. Finally, if we increased the weight still further we would reach a stage where the metal would fracture. This occurs when the external force is able to overcome totally the attractive forces between the atoms. Here we come to the second feature of the force-separation curve of Figure 11, namely, the point Q. If the external force acting on each atom is greater than f_q we are able to pull the atoms apart. With a bulk specimen this corresponds to fracture or rupture of the specimen.

Brittle Solids

There is a class of solids that merits a separate heading. These are materials which show practically no plastic deformation. If they are stretched they extend elastically and then snap. These materials are referred to as brittle solids. Typical examples are glass, porcelain, vulcanite, certain metals such as germanium and cast iron, and certain hard materials such as tungsten carbide and diamond. In these materials because of their structure and the nature of the bonds between the atoms it is easier to pull the atoms apart than to slide them over one another. This is the essential feature of brittleness.

We see from these simple examples that we can understand the basic deformation properties of a solid in terms of the forces between its atoms or molecules. In reality the situation is more complicated since the observed deformation properties depend not only on the force-separation curve but also on the presence of flaws, microcracks or dislocations in the arrays of atoms that go to make up the solid.

The Crystalline Structure of Solids

The atoms which compose a solid can be assembled in a number of ways. One way is that which occurs in pure metals where the atoms are in a regular array in an arrangement resembling the close packing of hard spheres.† For example, Figure 15a shows the arrangement of the atoms in copper or

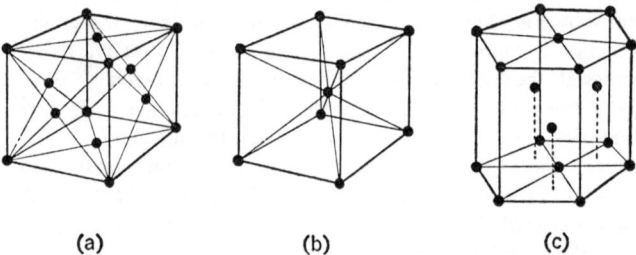

(a)   (b)   (c)

Figure 15   *Crystal structures. Arrangement of atoms in some typical metals: (a) copper or aluminum; (b) iron; (c) magnesium or zinc.*

aluminum; 15b, the arrangement in iron; and 15c, the arrangement in magnesium or zinc. The main feature is that the arrangement is regular or periodic and can be repeated over and over again with the same regularity. This is known as a crystalline structure. It is possible to grow single crystals of pure metals which are several centimeters in size. This means that millions of atoms are arranged in regular array in three dimensions, the arrangement of one portion being in exact register (almost) with every other portion. We have in-

† Strictly speaking the atoms in a metal ought not to be called atoms. Each atom gives up one of its more loosely bound electrons so that it is left with a positive charge. The electrons are distributed throughout the specimen. One may more accurately regard the metal as consisting of closely packed positive "atoms" in a sea of electrons. These electrons are easily displaced if a voltage is applied across the metal and in this way they enable the metal to conduct electricity.

Some Properties of Solids—and of Liquids

serted the word "almost" because in practice even the single crystal contains defects known as dislocations, or vacancies or occasional atomic impurities. Most industrial metals, say, a bar of copper or aluminum, are composed of hundreds of thousands of micro-crystals (see Plate Ia); the atoms in the individual crystals are in regular array but they are randomly arranged relative to their neighbors so that there is a certain amount of atomic mismatch at the crystal or grain boundaries (see Plate Ib). Commercial metals also usually contain quantities of impurities: these may settle out at the grain boundaries and greatly influence the bulk properties of the metal. Again many industrial metals are alloys consisting of two or more metals (or other elements) in intimate combination. Steel is a complex material consisting of iron, carbon, manganese and often chromium and cobalt.

Many solids, not only metals, have a crystalline structure. Typical examples are sodium chloride (table salt), diamond, camphor, sugar and numerous minerals and gem stones. Some minerals have an extremely complicated crystalline structure.

There is one aspect of the crystalline state that we may mention. Because of differences in distances between various atomic layers or differences in bond strengths, layers in one direction in the crystal may prove to be less tightly bound to the neighboring layers than layers in other directions. In that case we might expect plastic flow to occur more easily along these weakly bound directions. We might also expect fracture to occur more easily along these planes. This is often found to be so. Some crystalline solids have strength properties which vary so markedly with "crystallographic direction" that they will only fracture or cleave along a single crystallographic plane. This is true of mica. It also applies to graphite and to molybdenum disulphide and as we shall see in a later chapter this property has a direct influence on their frictional behavior.

The Free Surface

Before we go on to discuss other types of solids we may mention an important feature that is well illustrated by Figure 15. It concerns the difference between atoms in the surface and atoms in the bulk. If we take any atom in 15a well below the surface of a crystal we shall find that it has twelve near neighbors; the same applies to atoms in 15c; while in 15b each atom in the bulk has eight near neighbors. At the free surface, however, an atom has lost about half its proper quota of near neighbors. The surface atoms are clearly in a very different situation from those in the bulk.

First, they are most anxious to acquire further neighbors. This makes them highly reactive chemically. For example, a clean surface of pure copper if exposed to the air will grab hold of the atmospheric oxygen and react with it to form a thin layer of copper oxide in the twinkling of an eye. Conversely if we have a piece of copper already covered with a thin oxide film we shall find it very difficult indeed to remove the oxide and leave a truly clean surface. We shall have to heat it in a vacuum to a very high temperature until the surface layers evaporate away. This applies not only to crystalline solids but also to amorphous solids and even liquids (see below). The free surface is extremely reactive.

The second consequence is that whereas atoms in the bulk of the material are pulled equally in all directions by their neighbors, the surface atoms are pulled inward: the surface layer is in a state of stress and this produces what is known as "surface tension." Surface tension effects are not obvious with solids but they are with liquids. For example, surface tension pulls a small piece of liquid into a spherical shape. These forces are not big enough to pull a piece of copper into a spherical shape unless the temperature is raised until the metal is near or at its melting point. Again surface tension forces will draw a liquid up a narrow tube. Finally, surface tension factors determine whether a liquid will "wet" a solid and spread over the surface or remain as droplets. Roughly speaking if the surface tension of the liquid is less than the surface tension of the solid it will spread: otherwise it will collect together as individual droplets.

The Amorphous Structure of Solids

The word amorphous is of Greek origin and implies "without shape or form." It is used to describe non-crystalline struc-

Some Properties of Solids—and of Liquids

tures. A typical example is glass, which is a complex compound of silicon, oxygen, sodium, lead and other elements. A schematic illustration of quartz (a special type of glass which contains only silicon and oxygen) is shown in Figure 16, where the dark circles represent atoms of silicon and the open circles atoms of oxygen. The amorphous form is represented by 16a. If the quartz is kept at a high temperature for a long period it may rearrange itself and gradually crystallize (16b). With glass this process is known as devitrification, that is, ceasing to be glasslike.

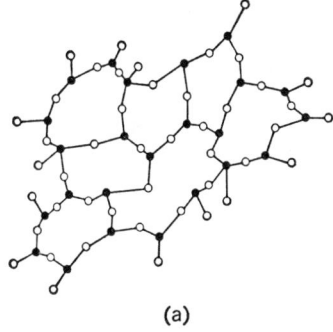

(a)

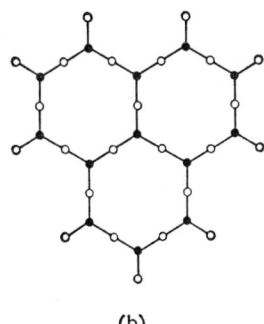

(b)

Figure 16 *Arrangement of atoms in quartz. Open circles represent oxygen, closed circles silicon. (a) Amorphous structure: glass. (b) Crystalline structure: devitrified glass.*

Some solids are made of such awkwardly shaped molecular units that they can never crystallize. This is true of some types of glass and of many plastics. For example, PVC (polyvinyl chloride) is a truly amorphous solid.

The Structure of Rubber

Both natural and synthetic rubber consist of long molecular chains of carbon and hydrogen. The chain may contain thousands of carbon-hydrogen units. In the latex which is extracted from the rubber tree the molecules are separate entities and the material is a liquid of very high viscosity (see below). The latex is converted into rubber by creating chemical links between the molecules at various points along the chain. Very often these links are formed by introducing sulphur; the sulphur atom can tie together the carbon atoms in two neighboring chains. These ties or bonds are usually referred to as cross-links.

In a piece of rubber in its normal state the chains are not arranged in neat parallel array. On the contrary, they are coiled up and tangled together in a manner shown schematically in Figure 17. In fact the average distance between the

Figure 17 *Structure of rubber. Long coiled-up chains of molecules are linked together by chemical bonds. The greater the number of cross-links the greater the resistance of the rubber to deformation.*

ends of a chain may be only a tenth or a hundredth of the stretched length of the chain. It is as though the rubber consists of a tangle of coiled-up springs. If we now attempt to extend the piece of rubber we pull on the coiled-up chains, and uncoil them. Clearly the rubber can be subjected to an enormous extension (ten to twenty times its original length) before the chains are pulled straight. In this condition the rubber now shows certain crystalline features because the chains are, more or less, parallel and close-packed.

For reasons which are well understood but which we cannot explain in simple terms, the resistance to deformation depends on the length of the segments between each link rather than on the length of the rubber molecule as a whole. The shorter the segment the greater the resistance to deformation. Consequently we can change the elastic properties of the rubber by varying the number of links. The more links we put in, the harder or stiffer the rubber. In the limiting case we may introduce so many links that the material ceases to be rubbery and becomes a hard solid like vulcanite or ebonite.

The detailed behavior of a rubber depends on the number of cross-links and on the structure of the molecular chains. It also depends very markedly on the temperature. As the temperature is reduced the coiled-up molecular springs get stiffer and stiffer. At very low temperatures the rubber becomes hard and stiff and even brittle. This raises problems for automobile tires operating in arctic conditions. In some rubbers the chains do not entangle with their neighbors and slip easily over one another, so that when the segments are uncoiled during deformation and then allowed to coil up again when the deforming force is removed, very little energy is lost. The rubber is then said to be very resilient. For example, if a ball of such a rubber is dropped on to a hard floor it will rebound almost to its original height. For this reason a resilient rubber is sometimes called "bouncy" or "lively."

By contrast some rubbers have a molecular structure with clumsy side groups attached to the main chain. During coiling and uncoiling of the segments, energy is expended in dragging these awkward groups past one another. A good deal of energy may be lost in this process and the rubber is said to have a low resilience: it will not bounce well and for this reason is sometimes called a "dead" or "soggy" rubber.

Another term that is used to describe the lack of resilience of a soggy rubber is "hysteresis loss." Hysteresis is a rather frightening word of Greek origin which conveys the idea that in loading and then unloading a specimen we do not end up where we started. We have lost something en route. Thus a rubber of high resilience has low hysteresis loss: a rubber of low resilience has a high hysteresis loss. This energy loss appears as heat in the rubber.

Most technological rubbers, for example, those in automobile tires, contain a good deal of "filler" consisting of very fine particles of carbon. This increases the hardness of the rubber and its resistance to wear. It also usually leads to an increase in the hysteresis loss properties of the rubber. Since carbon particles are much cheaper than rubber, a filled rubber is (or should be) appreciably cheaper than an unfilled rubber of the same weight.

The Structure of Plastics or Polymers

Most of the plastics that we use in everyday life, polythene, nylon, Dacron, Plexiglas, PVC (polyvinyl chloride), are long chain compounds of carbon, hydrogen and other atoms. The chains are not cross-linked so that they resemble latex; in fact if we melt nylon or Dacron it becomes a rather viscous liquid; it is because of this that these polymers when molten can be extruded through very fine nozzles to form the fibers that are used in synthetic fabrics.

At room temperature these materials are solid. If we take a polymer such as polythene and melt it and then cool it so that it slowly solidifies, parts of the polymer chains will pack together in a regular crystalline array but parts of the same chain will retain a tangled irregular structure and perhaps yet a further part will find itself incorporated into the crystalline portion of another chain (Figure 18). Polythene is, in fact, partly crystalline and partly amorphous. Some polymers such as PVC have an awkward molecular structure and can never pack to form even a partially crystalline solid: PVC is indeed always amorphous.

Some Properties of Solids—and of Liquids

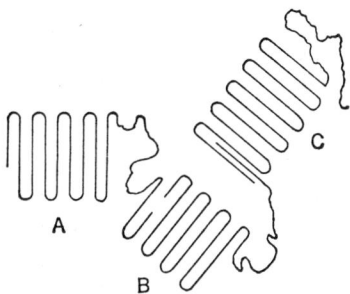

Figure 18 *Structure of a typical polymer such as polythene. The long chains can form crystalline portions over part of their length and coiled-up portions over other parts of their length. One molecular chain may become part of several crystallites.*

If we take a thin rod of plastic, suspend it vertically, and hang weights at the free end it will stretch elastically. Most of the deformation probably occurs in the amorphous part of the plastic. The resistance to elastic deformation resembles that of rubber but it would be more correct to say that it resembles rubber in which the coiled chains are relatively short and therefore rather more difficult to extend.

If we increase the weight beyond a certain value the plastic will begin to flow and the polymer chains will gradually become more aligned. A polymer such as polythene or nylon can be "drawn" in this way to three or four times its original length and by the end of the process it may become highly crystalline. The flow is not reversible: on removing the weight the polymer remains in its extended state so that, in some ways, the deformation resembles the plastic flow of metals. However, the flow of polymers increases very markedly with the time of application of the load. For this reason it is often considered to resemble the flow of an extremely viscous liquid (see below). Because polymers of this type show both elastic properties and viscous-like flow properties they are sometimes called visco-elastic materials.

There is a completely different type of polymer in which the long molecular chains are very strongly cross-linked. These materials are very hard, they do not flow and if they are hit hard they tend to fracture rather than deform. They can easily be molded and are often used in electrical fittings and in high-quality plastic crockery. Typical materials in this group are Bakelite, melamine, dexrin.

The Structure of Wood

Wood has enormous uses in furniture and in building construction. All woods consist of cellulose fibers which are packed together more or less parallel to one another and glued together by a substance called lignin (see Plate II). Some of the fibers contain short cross-fibers which reinforce the structure but this does not prevent the wood from having markedly directional properties. It is much easier to split wood parallel to the fibers than across them. The fibers are hollow and fluids can travel along them and so nourish the tree. The lignin and the fibers also contain waxes and resins and these play a very important part in determining the properties of the wood.

Balsa wood is extremely light: it is full of empty spaces and has a density which is only one fifth of that of water. Even when it is thoroughly waterlogged it will still float. By contrast lignum vitae (the name means "tree of life") has a density 1.3 times that of water: it will not float. It contains a great deal of lignin and is very rich in waxes and resins. It was discovered in the Caribbean islands by European travelers in the sixteenth century and was called the "tree of life" because the substances extracted from it (especially guaiacum) were supposed to have beneficial medicinal properties. Lignum vitae is a very tough wood. The fibers are arranged in a criss-cross array and the wood is difficult to split. In addition, as we shall see in Chapter VII, the waxes it contains make it an excellent low-friction material.

The Liquid State

We shall need to know something about liquids when we come to discuss lubrication. The simplest approach is to consider what happens if a crystalline solid is heated until it melts. We could, for example, consider the behavior of an ice crystal and the way in which the individual water molecules are able

Some Properties of Solids—and of Liquids

to break away from the rigid crystal structure and become "mobile" when melting is reached. However, it is more convenient to discuss an even simpler solid consisting only of atoms of one sort closely packed together, say, a single crystal of copper.

At room temperature the individual copper atoms are in their equilibrium positions as shown schematically in Figure 19a. The atoms, however, are not fixed rigidly but vibrate a

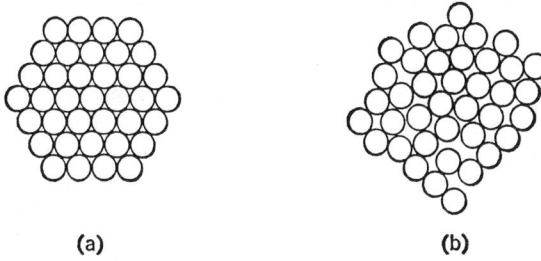

(a)   (b)

Figure 19 *Figure illustrating transition from solid to liquid state. (a) In a solid single crystal of copper there is a regular arrangement of atoms. (b) After melting, the ordered arrangement is broken down and there is more room between the atoms: this corresponds to the liquid state.*

little about these positions. In fact the amount of vibration is a measure of the temperature of the solid. As the temperature is raised the atoms acquire more "thermal energy": the vibrations become more vigorous; the amount of movement becomes larger. At some critical stage two marked changes occur. First, the ordered arrangement of the crystal is broken down and there is more room between the atoms. This is shown schematically in 19b. The second change is that, because the atoms have more energy and because there is more space between them, they are able to jump around and swap places. The material is now a liquid. If we apply a very small force we can make the atoms flow over one another. The liquid can, indeed, flow under a force immensely smaller than that required to produce plastic flow in the solid. As we have seen, this is simply because the thermal motion of the atoms has completely loosened up the structure and provides most of the energy necessary to overcome their interatomic attractions.

A finite force is required to make a liquid flow: the resistance to flow is called viscosity. Liquid copper has a very low viscosity. This is true of most liquids composed of very small atoms or molecules, since it is not difficult to pull one atom or molecule away from its neighbor and slide it into a new position. Water has a low viscosity. On the other hand a liquid consisting of very long molecules has a high viscosity. This is because far more work must be used in overcoming the attractive forces between large molecular units. Typical examples are lubricating oils or molten polymers.

There are four important features about viscous flow. First, the viscous resistance increases with the rate of deformation. If we wish to make a liquid flow fast we have to exert a greater force than if we make it flow slowly. Second, the molecules do not go back to their original position when the applied force is removed. The flow involves a non-reversible change and the work done in producing viscous flow appears as heat in the liquid. That is why the oil in an engine warms up after the engine has run for some time. The third feature is that the viscosity of a liquid becomes smaller as the temperature is raised. This is because there is more thermal energy enabling the molecules to escape from their neighbors: less external force is required to hurry them on. With a lubricating oil, for example, each rise in temperature of 10°C decreases the viscosity by a factor of about 2. Thus a rise of 20°C decreases the viscosity by a factor of 4, a rise of 30°C by a factor of 8, a rise of 40°C by a factor of 16. Similarly if we reduce the temperature by, say, 20°C we *increase* the viscosity of the oil by a factor of 4. This is why it is hard to start an engine on a cold morning: the oil is too stiff. As it warms up it becomes less viscous and the engine runs more easily.

The fourth feature concerns the effect of pressure. We can see from our model of the liquid state that if we squeeze the atoms or molecules closer together we make it harder for them to escape from their neighbors. This means that it becomes harder for the atoms or molecules to flow over one another: the viscosity increases. At modest pressures such as those that occur in the bearings of an automobile engine or in the bear-

Some Properties of Solids—and of Liquids 45

ings of a turbine the effect is not large. But there are some situations where the effect can be enormous. For example, the contact pressures between the teeth of gear wheels can be very large compared with the pressures occurring in an ordinary bearing. Under these conditions the viscosity of the oil may be increased a hundred thousand times. This is a very fortunate state of affairs, for it implies that the harder one tries to squeeze out the lubricant the higher its viscosity and the greater its resistance to extrusion. Clearly nature has, in this respect at least, been much kinder to the engineer than he might have expected. It is probable that many pieces of machinery function successfully in practice precisely because of this natural property of most lubricants to increase their viscosity when they are subjected to high pressures.

The Gaseous State

In a liquid the individual atoms (or molecules) have enough thermal energy to break away from one set of neighbors but they are recaptured by another set of neighbors. Only at the free surface of the liquid can they sometimes escape completely and this accounts for the presence of vapor above the liquid surface. If, however, we heat the liquid further we may so increase the thermal energy of the atoms that they are able to escape completely from all their neighbors. We can see this simply by considering once again the behavior of the two golf balls discussed earlier. If both balls are placed on a table and one is rolled slowly close to the other the attractive force may be large enough to bring the balls together and hold them in contact. Suppose we now roll the ball faster. It will again be attracted by the other and it may even collide with it: but because it has a high velocity (really a high energy) it will be able to break away and escape. This is the situation in a gas. The individual atoms (or molecules) may collide, they may stick together for an instant, but they break away and are able to lead an individual existence. They will go on colliding with other molecules and finally they will collide with the walls of the vessel in which the gas is contained. Here they will strike a molecule in the vessel wall and rebound into the bulk of the gas once more. The impact of the molecules on the walls of the vessel is the mechanism by which the gas exerts a pressure on its container. If we increase the temperature we increase the thermal energy of the molecules—that is, their velocity—so that they impart a greater impact on the vessel walls. Consequently the pressure is increased. Similarly if we reduce the volume of the vessel the molecules have to travel a smaller distance before they collide with the walls; the number of impacts per second is increased and this in turn implies an increase in pressure.

We see therefore that in a gas the individual molecules have so much thermal energy that they are able to escape from one another's company. The gas pressure is due to collisions of the molecules with the walls of the container. The pressure exerted by a given quantity of gas is increased if we raise the temperature or decrease the volume.

IV WHAT IS A SOLID SURFACE AND WHAT HAPPENS WHEN TWO SURFACES ARE PUT TOGETHER?

> Putting two solids together is rather like turning Switzerland upside down and standing it on Austria—the area of intimate contact will be small.
>
> F. P. Bowden,
> BBC Broadcast, 1950

The Importance of Surface Contours

It is clear from our discussion in Chapter II that we shall never understand how friction arises unless we know the way in which surfaces make contact when they are put together. Before we can do this we need to study the shape and contour of the surfaces so that we can form some idea of how the two surfaces fit together when they are placed on one another.

It is extremely difficult to prepare surfaces that are truly flat and smooth. This difficulty is partly a technological one. Consequently by improving techniques we can obtain smoother and flatter surfaces. However, part of the difficulty lies in the nature of the solids themselves. For example, metals used in engineering practice often consist of polycrystalline materials, the crystals often composed of different constituents and arranged in random orientations. Certain crystals will be harder than others or may point in orientations that are more difficult to deform. Consequently when such materials are polished parts of the surface will be removed more easily than other parts. Even with a single crystal of a pure metal an atomic plane, which in principle is atomically smooth, may contain defects which create locally soft regions that can be removed preferentially during polishing. If we try to grow a single crystal from the molten state, the surface planes will often contain steps, ledges and terraces several atomic layers high.

If we melt a piece of glass and allow it to solidify slowly, the surface may appear to be smooth. But because the structure of glass is not uniform (see Figure 16a) different parts of the surface will contract by different amounts on solidifying and the surface will acquire a waviness. If the surface is polished the lack of uniform structure will again imply that different parts will be removed more easily than others. The same problem arises with polymers. Wood is even more irregular in structure and a polished surface will show roughnesses which are very large on an atomic scale.

The only surface that we know of which can be molecularly smooth over relatively large areas is mica. This material has a single cleavage plane and layers can be peeled off to leave a surface which has no molecular steps over areas of several square inches.

Even the best engineering surfaces, prepared by grinding and polishing, have roughnesses of the order of a millionth of an inch. This seems very smooth in ordinary terms but one millionth of an inch is 250 Å or about 100 atomic spacings: clearly on an atomic scale the surfaces are rough. Most engineering surfaces are considerably rougher. How can we study the shape and contour of such a surface?

Stylus Method: The Profilometer

A very simple method is to pass a fine needle or stylus over the surface and measure its up-and-down movement by suitable sensitive devices. Such a method is used in the Talysurf Profilometer and in the Brush analyzer. The needle usually has a diamond or sapphire tip with a radius of about one ten thousandth of an inch. The vertical movement is magnified electrically and recorded on a pen recorder or on a suitable meter. Vertical magnifications of about 40,000 can be achieved so that the instrument can record surface features as

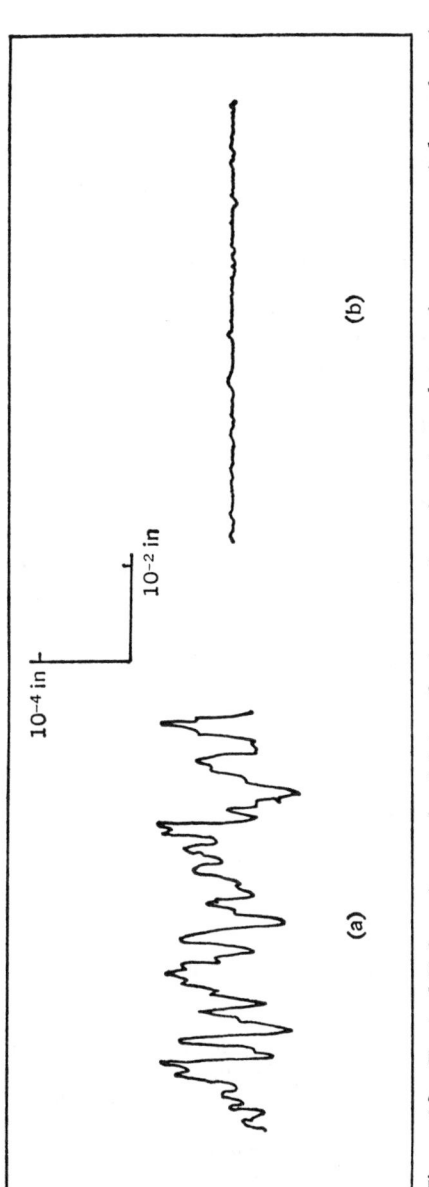

Figure 20 Typical Talysurf records of (a) a finely ground steel surface—the roughnesses are of the order of 50 microinches (millionths of an inch); (b) a very highly polished steel surface—the roughnesses are now only a few microinches.

small as one millionth of an inch in height. Its ultimate sensitivity is limited by the finite size of the needle, which prevents it from penetrating the finest scratches or irregularities. Typical records of a finely ground steel surface and a highly polished steel surface are shown in Figure 20.

Oblique or Taper Sectioning

Another method is to cut a section across the surface and examine the section in an optical microscope. For example, if there is a groove in the surface as in Figure 21a the section shows the groove ABC as in Figure 21b. This type of sectioning is commonly used by metallurgists and it reveals not only

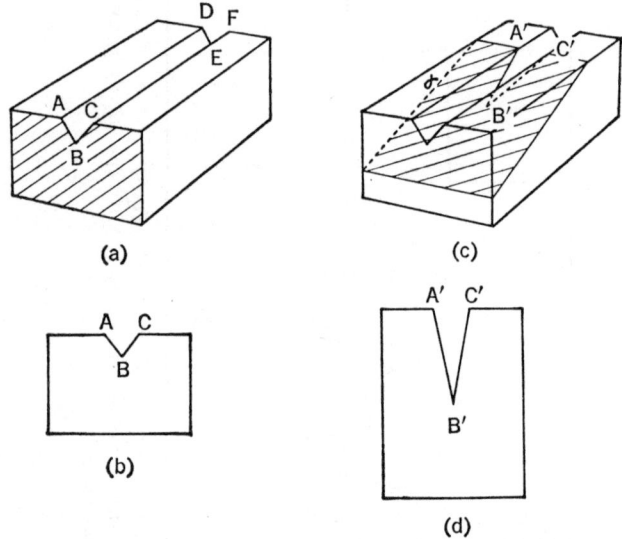

Figure 21 (a) *Examination of a groove ABC, DEF in a surface.* (b) *A section at right angles to the surface reveals the groove as ABC with equal horizontal and vertical magnification.* (c) *An oblique or taper section making an angle α to the surface magnifies the vertical scale compared with the horizontal; the groove is shown as A'B'C'. For $\alpha \cong 6°$ the vertical magnification is ten times bigger than the horizontal.*

Solid Surfaces and How They Make Contact 51

the surface shape but also the structure of the material immediately below the surface. If we wish to examine the surface shapes in greater detail we can do so by cutting the surface at a small angle α to the surface. The groove then acquires a shape shown in Figure 21c. If the angle $\alpha = 6°$ the height of the groove is increased by a factor 10 relative to the horizontal dimensions. (The vertical magnification relative to the horizontal is, in fact, equal to the cotangent of α.) Thus taper or oblique sections of this type reveal much more surface detail than the conventional metallographic section. However, they only reveal those grooves which are cut by the section itself. If the specimen is ground or polished in a single direction this is all the information we need. If, however, it is polished randomly so that the grooves run in all directions several sections in various directions must be made to obtain a comprehensive picture of the surface.

There are two points that should be made concerning taper or oblique sections. First, the free surface must be protected before cutting the section. This is achieved by electroplating the surface with a layer of metal of approximately the same hardness. Secondly, great care must be taken in polishing the section so that the polishing process itself does not introduce artificial features in the contour. Some typical results are shown in Plate III for metal surfaces prepared in various ways. If we can protect the surface adequately and prepare a polished section this technique can also be used for nonmetals.

Optical Methods

If we examine a surface with an optical microscope we find that we cannot distinguish details smaller than about 1000 Å across. However much we may enlarge the resultant picture we shall never be able to extract details from it smaller than this amount. This is known as the "resolving power" of the microscope and the limitation arises from the fact that the wavelength of light is itself too large. We can improve on this, in relation to surface contours, if we examine a taper section cut at an angle $\alpha = 6°$. The microscope itself will, as mentioned above, only be able to resolve the surface shapes to a limit of 1000 Å. But since the taper section itself magnifies the height of the surface irregularities by a factor of 10 this means that we are able to resolve surface roughnesses, the true heights of which are only about 200 Å.

There is also a special method of studying the height of surface irregularities without the need of preparing special sections. The surface is examined directly, using what is known as the principle of optical interference. Since we shall not quote examples using this technique we shall not describe it further except to say that it can resolve height differences as small as 50 Å. Interference microscopes are often used to study the smoothness and shape of accurate machine parts.

Electron Microscope

In some ways the electron microscope resembles the optical microscope: an electron beam replaces the light source and electrostatic or magnetic lenses replace the optical lenses (Figure 22a). The electrons usually have an energy of 50,000 to 100,000 volts. The electron beam may be directed through the specimen being examined and the transmitted electrons focused to form an image on a fluorescent screen or a photographic plate. This is known as transmission electron microscopy. Since electrons cannot penetrate great thicknesses of material the transmission method can only be used if the specimen is very thin. Sometimes it is possible to strip a thin surface film off the specimen and examine it in this way. A more usual method is to form a thin replica of the surface, using a plastic material such as Formvar, and to examine the replica by transmission in the electron beam. When suitable precautions are taken it is possible to obtain a resolution of about 5 Å so that it is almost possible to see individual atoms. A typical result obtained from the replica of a polished aluminum surface is shown in Plate IVa. The surface is covered with little hills and valleys about 100 Å high. Similar results obtained from a surface of electroplated gold are shown in Plate IVb; the hills are about 1000 Å high.

Another method of examining the surface without making a replica is to tilt the electron beam so that it strikes the specimen at glancing incidence (see Figure 22b) and then focus

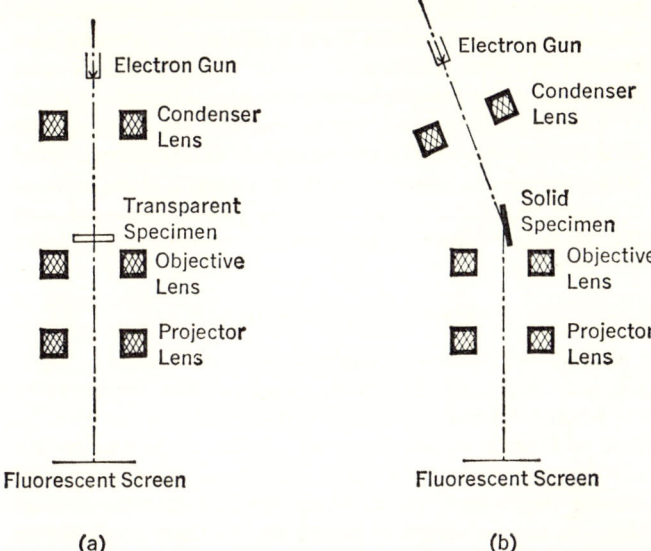

Figure 22 (*a*) *Electron microscope working in transmission. The specimen must be very thin so that the electron beam can pass through it.* (*b*) *The electron microscope working in reflection: the surface of the specimen is examined directly by the beam.*

the scattered electrons in the usual way. This is known as reflection electron microscopy. The resulting image provides a picture of the surface irregularities somewhat in the same way as a pedestrian sees the irregularities on a road surface illuminated by the headlights of an oncoming automobile. Long shadows are cast but these are foreshortened by the low angle of viewing. A typical example of the cleavage steps of mica is shown in Plate Va. (The cleavage was deliberately carried out clumsily so as to show a large number of these steps.)

In a modification of the reflection method a fine pencil of electrons scans the surface in a very tightly packed zigzag (in television jargon this scanning pattern is known as a "raster"). The electrons scattered by the specimen are picked up by a collector and used to modulate the brightness of the beam in a television tube; this beam scans the television screen in synchronism with the electron beam scanning the specimen (see Plate VIa). This technique is known as the Microprobe or Stereoscan and although the resolving power is not as good as in the other electron microscope methods the resulting pictures have a remarkably three-dimensional appearance. Plate VIb, for example, shows a needle scratching a metal surface: it reveals the crystal grains on the surface as well as the deformation and furrowing produced by the needle.

The Shape of Surface Roughnesses

All stylus profilometer methods and most sectioning methods are deliberately devised to exaggerate the vertical scale. As a result ground or polished surfaces tend to look like Alpine or Himalayan landscapes. If, in fact, we reconvert these surface profiles into their true perspective we find that the surface roughnesses are much gentler and that the average slope of any rising hill or falling valley is not much greater than a few degrees, perhaps 5°. This applies even to relatively rough surfaces such as those produced by a coarse file: the grooves are deeper but the average slopes are still only a few degrees.

There is of course one marked exception. With carefully annealed specimens of crystalline material the individual grains may project undisturbed to the very outermost surface. In this case the face of the crystal grain may make a very large angle with the surface. For example, with a solid possessing cubic structure, steps making 90° with the general level may occur. With ground or polished surfaces and with electrolytically polished surfaces this will not be the case except perhaps for the top two or three atoms. The fine-scale geometric structure, as distinct from the atomic structure, will consist of gentle slopes. It is more accurate, if perhaps less picturesque, to describe surfaces as resembling rolling downs rather than mountain peaks.

The Contact Between Surfaces

All these techniques show that surfaces, however smooth

Solid Surfaces and How They Make Contact 55

we may prepare them, are rough on an atomic scale. Placing two such surfaces together is rather like putting New Hampshire on top of Maine. Contact occurs only at the tips of the promontories. Over the rest of the surfaces there may be gaps of a hundred angstroms or more. Now, as we saw in the preceding chapter, atomic forces operate over very small distances of the order of a few atomic diameters. Consequently these gaps, in effect, completely separate the surfaces and they do not interact with one another. The load is therefore carried on the tips of the surface asperities.

We now restrict our discussion to the behavior of metal surfaces and in a later chapter we shall extend our ideas to the behavior of non-metals. As we have seen the surfaces first touch at the tips of the surface roughnesses. For minute loads the tips will deform elastically. However, as soon as the load exceeds a very small value (a fraction of a gram per asperity) the "elastic limit" of the metal is exceeded and plastic flow, that is, permanent deformation, occurs. The promontories yield plastically until the area of true contact is large enough to support the applied load.* It turns out that the pressure which metals can support when they undergo local plastic deformation does not depend greatly on the shape or size of the asperities. This flow pressure or yield pressure P is a material property of the metal. It is defined as the load W divided by the area of true contact A, that is to say,

$$P = W/A.$$

Therefore if we double the load W the area of contact A must be doubled in order to maintain a constant yield pressure. The area of real contact will not depend on the number of points

---

* Recent studies suggest that with extremely smooth surfaces the deformation of the outermost promontories may be elastic. Surfaces of such smoothness will be rare in engineering practice. In any case, for clean surfaces the adhesion and subsequent breaking of the junctions during sliding will lead to plastic flow. The newer results are more relevant to well-run-in lubricated surfaces. They change the details of the sliding process but not the main ideas discussed in this book. See, however, the exceptional behavior of mica, described in Chapter V.

of contact since the final pressure must always finally come out equal to P.

How large is P and what is the real area of contact? Experiments show that P is comparable with a quantity that metallurgists refer to as the indentation hardness of the metal. It has values of the following order of magnitude:

For lead $P = 500$ kg. per cm.^{2}
copper $P = 8,000$ kg. per cm.^{2}
mild steel $P = 12,000$ kg. per cm.^{2}
hard steel $P = 80,000$ kg. per cm.^{2}
tungsten carbide $P = 200,000$ kg. per cm.^{2}

What does this mean in terms of the area of real contact? Suppose we take two flat mild steel surfaces each of one sq. cm. and place them in contact with a load W. Plastic flow of the tips will occur until the total area A of real contact over all the tips is sufficient to support the load and the true contact pressure is P. As we saw above this occurs when

$$P = W/A$$
$$\text{or} \quad A = W/P.$$

If the load is, say, 100 kg. (a tenth of a ton) the area of real contact is less than one hundredth of a square centimeter, i.e., only one hundredth of the area of the surface is really in contact. If the steel surfaces are ten times bigger the points of contact will be distributed over the larger area and there may be more of them. But if there are more contact points each will have to carry a smaller portion of the load. The amount of plastic flow of each contact will be less. However, the sum of all the contact areas will again be given by $A = P/W$ so that we shall again end up with the same true contact area as before. The true area is now only one thousandth of the area of the surfaces. We see that the area of true contact may be a small fraction of the apparent area.

This conclusion has been supported by measurements of the electrical resistance across the surfaces. There are many difficulties, chief of which is that oxide films present on metal surfaces are electrical insulators. However, if appropriate pre-

Solid Surfaces and How They Make Contact

cautions are taken we can obtain reproducible results. They show once again that the load and the yield pressure of the metal determine the real area of contact. It depends very little on the roughness of the surfaces; nor does it depend on their apparent area.

The Nature of Metallic Surfaces Used in Engineering

Most engineering surfaces are prepared by turning or milling and finished by grinding or polishing. The material in the surface is heavily deformed and as a result the surface layer is often harder than the material well below the surface. On top of this deformed material, if the metals are reactive, a layer of metal oxide is formed and this in turn will be covered with adsorbed molecules of oxygen and water vapor. If in addition polishing has been carried out at high speeds (see Chapter VII) the surface layer may consist of a smeared "fudge" of metal, metal oxide and polishing powder. On top of this layer a new oxide will form, while the material below the surface will show traces of deformation produced in earlier stages of abrasion. The thicknesses of the various layers proceeding from the bulk of the metal out to the surface are approximately as follows:

Heavy subsurface deformation	50,000 Å =	200	millionths of an inch		
Intense surface deformation	10,000 Å =	40	"	" "	"
Polish layer	1,000 Å =	4	"	" "	"
Oxide layer	200 Å =	1	"	" "	"
Adsorbed gases	10 Å =	$\frac{1}{25}$	"	" "	"
Surface waviness	500 Å =	2	"	" "	"

Figure 23 shows a diagrammatic representation of some of these features. Such a surface is neither smooth nor simple. When two such surfaces are placed together (Figure 24) the individual contact zones are generally large compared with the thickness of the outermost oxide layers. Consequently the deformation is determined by the bulk properties of the metal at and below the surface. Whether the oxide film is penetrated or not is another matter. The point we wish to emphasize is

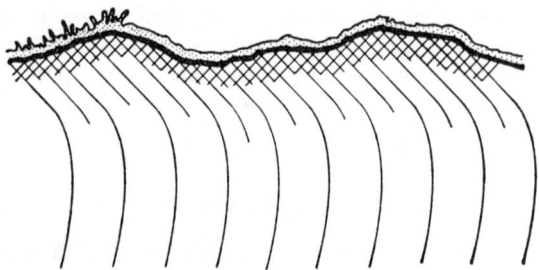

≈ Oxide about 1 microinch
▬ Polish or beilby layer 4 microinches
✕✕✕ Severe deformation 40 microinches
\\\ Heavy deformation 200 microinches
\ \ Minor deformation

Figure 23 *A typical surface after abrading and polishing in air. There is a heavily deformed subsurface layer 200 millionths of an inch (200 μ in.) thick, an intensely deformed surface layer 40 μ in. thick, then a polish layer 4 μ in. thick, then an oxide film 1 μ in. thick and on top of this adsorbed gas and vapors perhaps $\frac{1}{25}$ μ in. thick (1 μ in. = 1 microinch = 1 millionth of an inch).*

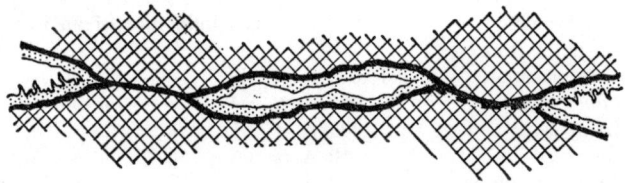

Figure 24 *Two surfaces such as those shown in the previous figure placed together make contact over areas much larger than the thickness of the outermost surface films.*

that the area supporting the load is determined by the average plastic properties of the metal itself near the surface. For this reason, as we saw above, the *real area of contact is, to a first approximation, proportional to the load and independent of the size of the bodies.*

As we saw in Chapter II the friction between two bodies is also proportional to the load and independent of the size of the bodies. Clearly there is a very close parallel between the friction and the area of true contact.

V FRICTION AND ADHESION OF METALS

> ... the flat Surfaces of metals or other Bodies may be so far polish'd as to increase Friction and this is a mechanical Paradox: but the reason will appear when we consider that the Attraction of Cohesion becomes sensible as we bring the Surfaces of Bodies nearer and nearer to Contact.
> J. T. Desaguliers (1683–1744),
> *A Course of Experimental Philosophy*
> (London, 1734), p. 182

Adhesion and Plowing in Friction

In the previous chapter we considered what happens when two metal surfaces are placed together under a normal load. ("Normal" here means that the load is at right angles to the surface of contact, which for brevity is often called the interface.) We showed that they make contact only at the tips of their roughnesses and that these are the regions that support the load. Here the atoms on one surface are in close proximity to those on the other, so that they are within the range of their very strong attractive forces. Consequently if we wish to separate the surfaces we shall have to overcome these forces: we shall have to expend a considerable amount of work in pulling or sliding the surfaces apart.

This conclusion applies whatever the nature of the solids since all atoms exert attractive forces on other atoms. In the case of metals, however, we can also describe the situation in large-scale terms. If, for example, we press two clean pieces of gold together they will virtually become a single piece across the regions of real contact. There will, in fact, be strong adhesion. This process, which is well known with metals, is sometimes called "cold welding," since the surfaces stick together strongly without the application of heat. Of course, if the surfaces are not clean the adhesion may be weaker, but it may still be appreciable.

If we now apply a tangential, i.e., a sideways force to one of our surfaces the junctions formed at the regions of real contact will have to be sheared if sliding is to take place. The force to do this is the frictional force. If the force we apply is smaller than this we shall deform the junctions but we shall not break them: sliding will not occur.

There is another part of the force necessary to produce sliding which may arise from a completely different mechanism. If one of the surfaces is much harder than the other it may plow out a groove (or the individual asperities on it may plow out a number of fine grooves) in the softer surface. The force to do this will contribute an additional "grooving" or "plowing" term to the friction.

These two mechanisms are illustrated by the taper sections shown in Plate VII. Plate VIIa shows the results obtained when a curved copper slider slides over a clean flat steel surface at a slow speed, at room temperature. It is seen that tiny particles of copper are left strongly attached to the steel. Clearly strong adhesion and shearing of junctions has occurred at the interface. In some cases the attachment is so strong that small fragments of steel have actually been plucked out of the steel surface. Plate VIIb shows the corresponding results obtained when a curved steel slider slides over a flat copper surface. The shearing of friction junctions is not very evident but the grooving of the copper surface and the heavy deformation of the copper below the surface of the track are clearly revealed.

These two taper sections summarize the frictional process for metals. *Friction is the force required to shear intermetallic junctions* plus *the force required to plow the surface of the softer metal by asperities on the harder surface.* This plowing term becomes important if one drags a needle over,

Friction and Adhesion of Metals 63

say, a lead surface; or if one rubs a file or pad of emery paper over a softer metal. In most unlubricated sliding systems the plowing term is small compared with the adhesion term and we may ignore it. Consequently the friction is essentially the force required to shear the intermetallic junctions.

If the true area of contact is A and if the force required to shear a unit area of intermetallic junctions is s, the force required to shear the junctions, which we equate to the frictional force, is simply

$$F = As.$$

Since A is proportional to the load and independent of the size of the bodies, it is clear that, if s is a constant for any given pair of surfaces, the friction will also be proportional to the load and independent of the size of the bodies. These are the two basic laws of friction. If, of course, the area of contact did not follow such simple laws, the laws of friction would also be more complicated. This occurs with polymers and rubber, where the frictional behavior deviates appreciably from Coulomb's or Amontons' laws; while with molecularly smooth surfaces such as mica, where the area of contact follows completely different laws, the frictional behavior bears no resemblance to the laws of Amontons.

Friction can be greatly influenced by s, the strength of the junctions. Any sort of contamination such as metal oxides or grease films will weaken the junctions and reduce s. This is why friction depends so critically on surface cleanliness. We might find a lubricant film that is so effective that it completely prevents the asperities from coming into intimate metallic contact. In that case no metallic junctions will be formed. The friction will fall by a very large factor; it will not, however, fall to zero since even if there were no metallic junctions we would still have to shear the lubricant film that remains between the surfaces if they are to be prevented from adhering.

The Evidence for Friction Junctions

Our argument for the formation and shearing of friction junctions is partly theoretical and partly based on the taper sections shown in Plate VII. They show that strong interfacial adhesion can occur even at room temperature. Another method which is far more delicate involves the use of radioactive tracers. If, for example, the slider is radioactive and is rubbed over a non-radioactive metal, the transferred fragments, which of course retain their radioactivity, may be studied by their radioactive emanations. They may be detected with a Geiger counter. A far more graphic method is to let the transferred fragments photograph themselves. A photographic plate is placed with its emulsion side in contact with the friction track so that the radioactive fragments sensitize the emulsion: the plate is then developed. The blackening produced gives a picture of the distribution of the fragments while the size and intensity of the blackened spots give a measure of their size. This technique has been given the impressive name of "autoradiography." Plate VIIIa is a typical autoradiograph of lead sliding over steel on clean and lubricated surfaces, showing the transferred fragments. With the lubricated surfaces shown in VIIIb, the lubricant has reduced but not entirely eliminated metallic pickup.

Adhesion of Metals

Although the taper sections and the autoradiographs show that there is marked adhesion between sliding metals there still remains something in this explanation of the mechanism of friction that is not entirely satisfactory. If adhesion occurs at the regions of contact, why in general do we not find a strong normal adhesion between, say, a piece of copper and a piece of steel when one surface is pressed on to the other? Surely we should expect that quite a large force would be needed to pull the surfaces apart in a normal direction, that is, in a direction perpendicular to the interface. This is not generally observed for two reasons. First, in applying a tangential force to initiate or maintain sliding there is more chance of breaking through surface films of grease, oxide or dirt so that junctions formed during sliding may be stronger than those formed under normal loading where no tangential force is applied. Indeed if experiments are carried out in a high vacuum with thoroughly clean metal surfaces pressed

Friction and Adhesion of Metals 65

together we do find some adhesion, but it is not as large as we would expect. Second, we always measure friction while the normal load (which presses the surfaces together) is retained. On the other hand, if we want to measure adhesion we must first remove the normal load. Although the points of contact are deformed plastically, the material a small distance away from the interface is deformed elastically; these deformations are reversible. When we remove the normal load the stresses in the elastic "hinterland" will be released and there will be a slight change in the shape of the interface. The junctions are heavily deformed plastically and are not very ductile: a small tensile stress will snap them apart. Consequently as we remove the normal load the junctions snap apart one by one so that by the time we are ready to measure the adhesion there are practically no junctions left (see Figure 25).

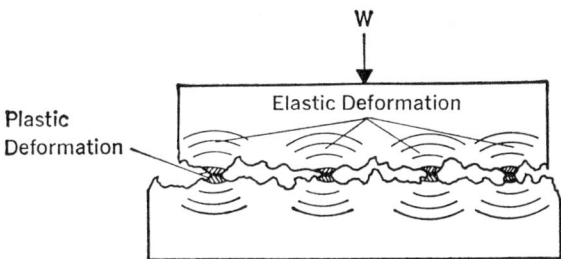

Figure 25 *Sketch representing contact between two surfaces. Plastic deformation occurs at the regions of real contact and here friction junctions are formed. The surrounding regions are deformed elastically so that when the load is removed these elastic stresses are released—this breaks the junctions.*

Initially this seems a dishonest argument: for we are saying that the junctions are there when we need them to explain friction but are not there when we wish to explain away the absence of normal adhesion. There is, however, a simple way in which we can show that this argument is perfectly valid. Suppose we carry out these adhesion experiments with metals for which the junctions remain soft and ductile even though they have been heavily deformed during their formation.

When we remove the joining load these junctions should be able to stretch in a ductile manner and survive the small changes in shape which occur at the interface. The adhesion should then be strong. This is exactly what happens. If we prepare a soft metal such as lead with a reasonably clean surface and press it on to another clean surface we will observe strong adhesion. The force to pull the surfaces apart may be as large as the original joining force and we will find pieces of lead adhering to the other surface. Desaguliers first described this in 1734. Even with harder higher-melting-point metals such as copper and aluminum the action of pressure can produce strong adhesion with clean surfaces at room temperature. In general, however, the adhesion occurs much more readily at higher temperatures, where the junctions remain more ductile. This is probably very important in the technological process of sintering. In this process very fine particles of metal are squeezed together and then heated. Though the sintering temperature is generally far below the melting point of the metal very strong adhesion occurs between the particles and the "compact" can be very strong indeed even if it remains partly porous. Sintering provides a convenient technique for fabricating metal rods, bars and tubes. It also has wide applications to certain types of non-metallic solids which can be prepared as fine particles but not easily in bulk form.

Adhesion and Friction

In the laboratory it is convenient to study adhesion by using the metal indium, which is only one quarter as hard as lead; it has a melting point of 150°C, and is much less reactive chemically than lead, so that its surface is not as readily contaminated with oxide films. If a hard sphere of some other material is pressed into the surface the adhesion will depend on the nature of the material. For a variety of metals such as copper, iron, platinum, and for non-metals such as diamond and rock salt the adhesion will be as large as the normal joining force. On the other hand, the adhesion will be much weaker if we use a "non-stick" material such as the polymer Teflon (PTFE) or if we deliberately deposit

Friction and Adhesion of Metals   67

a single molecular layer of lubricant on to the surfaces. Experiments of this type bring out the close connection between friction and adhesion. Friction is the shear strength of the junctions formed at the regions of real contact, adhesion is the tensile strength. Materials that give a low coefficient of friction generally give poor adhesion. High friction materials, on the other hand, will give, in principle, a strong adhesion. If there is poor adhesion it is due either to the effect of surface films or to the effect of released stresses in the elastic "hinterland."

These two factors also play a very important part in the action of practical adhesives. For glues and adhesives to function properly they must not only "wet" the surfaces thoroughly; first, contaminant films must be removed and, second, undesirable stresses in the material surrounding the glue or in the glue itself must be kept as small as possible.

The Wringing of Block Gauges

Readers who are familiar with standard block gauges used in engineering as precision standards of length are aware that these gauges can be "wrung" together to give an extended range of standard lengths. They may ask whether this type of adhesion, which can be very strong, is of the same nature as the inter-metallic adhesion we have just discussed. The answer is no. If gauges are thoroughly degreased and "wrung" together, metallic adhesion will indeed occur: junctions will be formed across the interface and the surfaces will be torn and roughened when they are separated. In fact the gauges will be ruined. In practice the gauges must be covered with a thin smear of oil or grease before they are "wrung" together. The adhesion is due to the viscous and surface tension forces exerted by the grease film itself.

Static Friction and Kinetic Friction

As mentioned in Chapter I, static friction is generally larger than kinetic friction. Why is this so? There are at least three mechanisms which may account for this. First, if the surfaces are at rest the area of contact may actually increase as a result of creep. (This effect is particularly marked with polymers.) Consequently the junction area to be sheared will be larger if the surfaces are stationary than if they are moving over one another. Second, with increasing time of contact diffusion of atoms from one solid, across the interface, to the other solid will be favored; this will lead to a strengthening of the junctions with time of contact.

There is a third mechanism which may be important if the surfaces are covered with a contaminant film. It is possible that a time factor is involved in the breakdown and penetration of such a film. The deformation of static contact is relatively gentle and does not easily disrupt surface films. If, however, a tangential force is applied, even though this is not enough to start sliding (i.e., less than the static friction), the stress situation at the interface favors breakdown of the surface film with a consequent increase in friction. On the other hand, this process of breakdown and penetration will occur less effectively if the surfaces are only in transitory contact, which is the situation once sliding occurs. Consequently the friction is lower. Thus the kinetic friction will be lower than the static. This mechanism is especially likely with lubricated surfaces: stationary surfaces may well penetrate the lubricant film whereas moving surfaces may slide over it.

Stick-slip Motion

In any mechanism where the kinetic friction is less than the static there will be a tendency for the motion to be intermittent rather than smooth. The surfaces will stick together until the sliding force reaches the static friction value. The surfaces will then slip over one another at the lower kinetic friction until once again they stick together.

We may explain this in terms of a highly simplified model. Suppose we rest body A on a flat surface (Figure 26). Then we attach a coiled spring to A and pull the free end of the spring B in a horizontal direction with a uniform speed of one unit of distance per second. We now wish to see if the body A will also move with the same uniform speed as B. At the beginning of the experiment (26a), before the spring

Friction and Adhesion of Metals

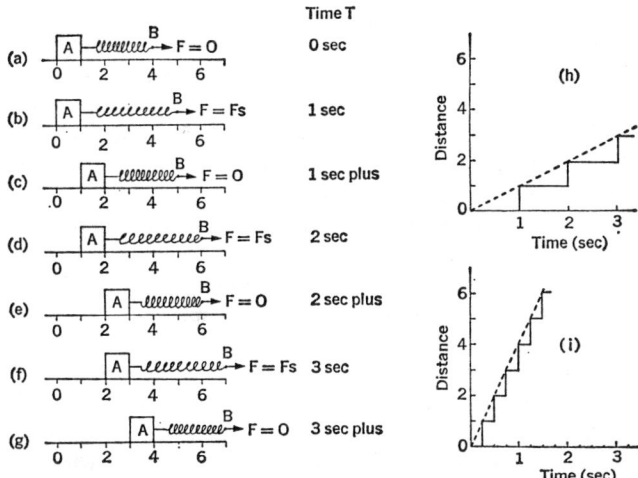

Figure 26   *A simple model showing that, if a body is pulled along by a spring, then even if the end B of the spring moves along at a uniform speed the body A moves along in jerks. This occurs because the static friction F_s is greater than the kinetic. The higher the speed of B the less jerky is the motion of A.*

has been stretched, i.e., just before B has begun to move, there is no tension in the spring (Time = 0). Let us assume that when the spring is stretched by one unit of distance the tension in the spring equals the static friction F_s. Then after one second the force that the spring exerts on A is F_s (Time = 1 sec., as in 26b). At this instant the static friction is exceeded and A begins to move. If the kinetic friction is very small compared with the static the spring is exerting far more force than is needed to make the body slide. It will slide very rapidly indeed, very much more rapidly than the speed of B. It will therefore restore the spring back to its unstretched length so that no force is being exerted on it. It therefore comes to rest.* We call this Time = 1 sec. plus

* Readers who are familiar with Newton's laws of motion will realize that because of its inertia body A will still continue to travel beyond this point. We can, however, retain the model described above by assuming that the motion of A is heavily "damped."

(Figure 26c). We shall now have to wait until B has moved another unit of distance (i.e., another second) until the spring has been stretched to the point where the tension is again F_s (Time = 2 sec., as in 26d). Sliding will then again occur and body A will shoot forward a further unit distance (Time = 2 sec. plus) as in 26e. In the third second the motion will be as shown in 26f and 26g.

If we plot the distance that A moves against time we obtain the result shown in 26h. It is seen that A has a very jerky motion. The dotted line shows the movement of B: this is a straight line corresponding to uniform speed. If we quadruple the speed at which B moves the jerks will be less marked (26i). If we make B move ten times faster the jerkiness of the movement of A is very small indeed.

The conclusion is: the jerkiness in the movement of body A is less marked the faster we pull the spring. This is what we should expect since the static friction operates for a very short time interval and the processes in the contact regions approach those operating during kinetic friction. A similar reduction will occur if we increase the stiffness of the spring so that a smaller extension is required to achieve a pulling force equal to F_s. There is a third factor which can reduce the size of the jerks, namely, the "damping" of the whole sliding system. Finally the jerkiness will be less marked if the static friction is little greater than the kinetic.

At this point the reader may well ask, "Who would dream of pulling a body with a flexible spring if he wanted to make it move with a uniform speed? Surely it would be more sensible to use a string, or better still a stiff rod." The answer is that even a stiff rod has a certain amount of elasticity and will stretch a little when it is pulled. This means that everything we have described above will still be valid but on a reduced scale. Elastic deformation occurs in all driving mechanisms. For example, the transmission shaft of an automobile will twist a little in transmitting torque from the engine to the rear axle. The teeth of gears will flex a little in transmitting force from one gear wheel to the other. A bicycle chain will stretch a little. A cutting tool on a lathe will deflect a little as it cuts away a chip of metal. All these examples indicate that if the static friction is higher than the kinetic, intermittent motion may occur. Indeed intermittent motion is of very common occurrence in sliding mechanisms and a great deal of ingenuity has been used in trying to overcome it, particularly in the operation of automated machinery. The squeaks and grunts generated by sliding surfaces usually arise from vibrations set up by the intermittent nature of the sliding process itself.

Of course intermittent motion may occur even when the friction does not fall with increasing velocity. For example, a hard slider rubbing on a soft surface may produce a pile-up of material ahead of it; this will increase the friction until some point of instability is reached when sliding suddenly occurs at the lower original value. Again, on a more practical level intermittent motion may occur in badly designed brakes. In effect, because of the way in which the brake is pivoted, the frictional force itself produces a large increase in the normal load so that a catastrophic increase in friction occurs: this is followed by a sudden yielding in some part of the mechanism with a subsequent drop in friction. This type of jerky motion is known as "spragging" and is essentially the result of poor design (see Figure 27).

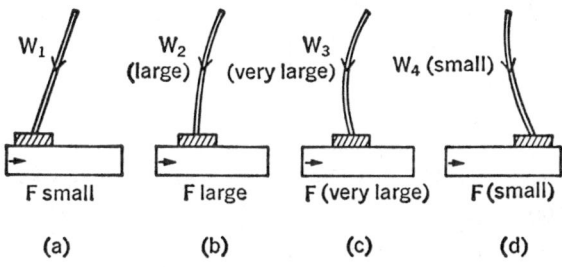

Figure 27 *The "spragging" behavior of a badly designed brake mechanism. The frictional force (a) bends the support and increases the normal force (b); this increases the frictional force and bends the support further until suddenly the support bends sharply and relieves the normal force (d). The friction now falls sharply.*

The Effect of Hardness on the Friction of Metals

Intuitively one might think that soft materials would always give a much higher friction than hard material, but the influence of hardness is, in fact, rather small. If, for example, we compare the friction of tin and the friction of hard steel, where the hardness varies by a factor of 100 or more, we find that the friction is only two or three times greater for the softer material. The reason for this relatively small difference in friction is illustrated in Figure 28. If one or both of the sliding metals is soft, the area of real contact A will be

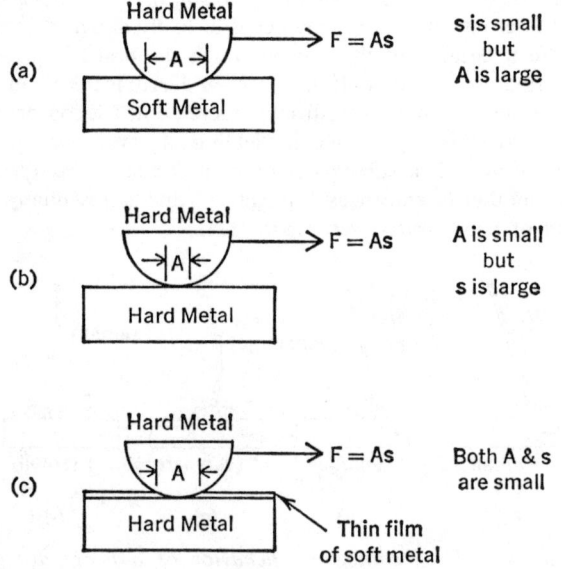

Figure 28 *Hardness and friction. The two top figures show why friction does not depend greatly on hardness. With a soft metal A is large but s is small: with a hard metal A is small but s is large. However, if a thin film of a soft metal (low shear strength s) is deposited on a hard metal (because it is hard the area of contact A is small) a low friction may be obtained since both A and s are small.*

Friction and Adhesion of Metals 73

large. However, the metal at the interface which has to be sheared when sliding occurs is a soft material for which s is small (28a). On the other hand, if we slide a hard metal on hard metal the area of contact A will be small, but s will be large. Consequently the product $F = A \times s$ is not very different in the two cases (28b). In fact, as we have just mentioned, the friction is higher for the softer metal, partly because the junctions can flow more readily, and partly because the larger deformation of the surface causes a more effective penetration of oxide film. The striking point, however, is not that soft metals have a higher friction than hard metals, but that on the whole the hardness has so little effect on friction.

How to Cheat the Laws of Friction

From what we have just said one would think that metal surfaces must always give approximately the same coefficient of friction. There is, however, one way in which we may obtain a low coefficient of friction of metal surfaces. The trick is to cover a hard surface with a thin film of much softer metal (see Figure 28c). The load is supported by the hard metal so that the area of contact A is very small. On the other hand, shearing occurs in the soft metal if we are careful and do not penetrate the metallic film. Consequently s is small. This enables us to obtain a situation in which both A and s are small and the resulting friction is relatively low. For example, for steel sliding on a flat steel surface $\mu = 0.8$. For steel sliding on indium the friction is about $\mu = 1.6$. If, however, the flat steel surface is covered with a thin film of indium the friction falls to a value of the order of $\mu = 0.1$. Of course if the loading is too severe the indium film may be penetrated and increased steel-steel interaction may occur.

Thin metallic films may be used as a means of reducing friction. They have been used effectively in the drawing of metal rods and tubes. This is a process in which a rod (or tube) of given diameter is pulled or "drawn" through a die to reduce its diameter (see Figure 29). A thin film of a soft metal deposited on the rod can greatly reduce the friction between the rod and the die. Recently thin metallic films have also found use as a lubricant for ball bearings operating in a

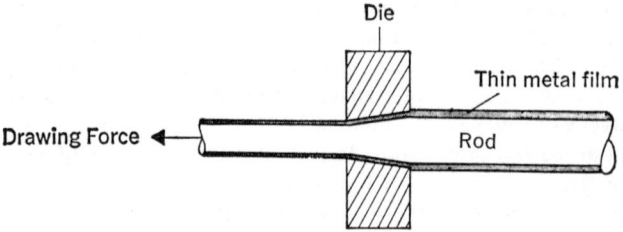

Figure 29 *A rod or tube may be reduced in diameter by "drawing" it through a suitably shaped die. The friction between the rod and die can be greatly reduced either by using a lubricant or by using a thin film of a soft metal.*

very high vacuum under conditions resembling those of outer space, where conventional lubricants would evaporate. It is surprising that although these films have no means of replenishing themselves they have shown quite an extensive life for this application.

Bearing Alloys

In most conventional machinery axles are run in bearings in the presence of lubricants (Figure 30). Provided the bearing is properly designed there should be little penetration of

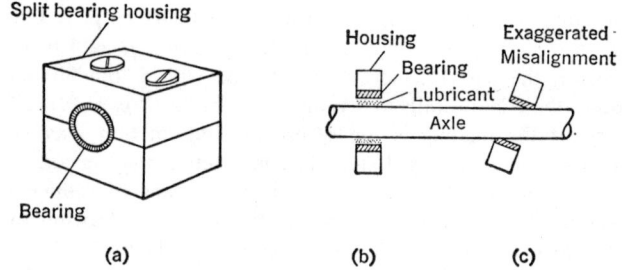

Figure 30 (*a*) *Typical bearing mounted in its "housing."* (*b*) *The bearing is so arranged that a film of lubricant separates the bearing from the axle.* (*c*) *If the bearing is not properly aligned it must be able to deform and take up the shape required by the axle* (*exaggerated misalignment*).

Friction and Adhesion of Metals

the lubricant film. Consequently the intrinsic frictional properties of the bearing may not be of great importance. Usually it is the mechanical and physical properties of the bearing that are far more important. The bearings must not be too hard, since they must be able to deform and take up the shape required by the axle, particularly if the axle and bearings are not properly aligned (30b). They must also have good heat-conducting properties so that any frictional heat can be readily conducted away. Many bearing alloys also contain a constituent with a low melting point so that if there is a danger of excessive heating local melting could occur and the molten film would be smeared over the surface. In this way seizure is prevented.

The oldest type of alloys used in bearings, the "white metal" alloys, consisted mainly of lead or tin; these have all the properties described above. Recent work, however, suggests that one of the most important reasons for their effectiveness as bearings in automobile engines has little to do with these properties. It turns out that many failures in automobile bearings arise from a piece of metal swarf (metal fragment left over from some cutting, milling or grinding operation) carried through the oil ducts from some other part of the engine. If this gets trapped between the axle and the bearing it may cause very heavy scoring and subsequent seizure. If, however, the bearing is relatively soft, the hard metal particle can be buried in the bearing so that it does not project above the rubbing surface.

Another class of bearing material which has been much used in aircraft is the copper-lead bearing alloy. This consists of a hard copper base with fine particles of lead dispersed through it. When sliding takes place the lead is smeared over the surface and the behavior is like that of a hard metal covered with a thin film of a softer metal. Although this is a low-friction type of material, it cannot be used effectively unless the bearings are well designed and adequate lubrication is provided. Since the bearing material is hard the alignment of bearing and axle must also be very good.

Some Typical Friction Values

We give in the following table some typical static friction values for metals cleaned in air so as to be grease-free. As we shall see in a later chapter, if the surfaces are thoroughly cleaned in a high vacuum the friction can be very much greater.

TABLE I
FRICTION VALUES FOR METALS IN AIR

Metal	μ_s
(a) Sliding on themselves	
Aluminum	1.5
Copper	1.5
Copper (oxide film not penetrated)	0.5
Gold	2.5
Iron	1.2
Platinum	3
Silver	1.5
Steel (mild steel)	0.8
Steel (tool steel)	0.4
(b) Sliding on hard steel	
Aluminum bronze	0.5
Brass	0.4
Cast iron	0.2 to 0.4
Copper	0.8
Copper-lead alloy	0.2
Phosphor bronze	0.35
Steel (mild steel)	0.6
White metal bearing alloy	0.6

VI FRICTION OF NON-METALS

... it is necessary to know the nature of the contact which this weight has with the smooth surface where it produces friction by its movement, because different bodies have different kinds of friction. . . .
Leonardo da Vinci

In this chapter we shall consider the friction of a number of non-metallic solids. Here again the friction is due to interfacial adhesion but the details are often strikingly different from those applying to metals.

Polymers and Plastics

With polymers such as polythene, Plexiglas (Perspex), polystyrene and nylon, the coefficient of friction is generally lower than for metals. With clean metals in air the friction, say for copper on steel, is of the order $\mu = 1$. With most plastics it is of the order of $\mu = 0.4$.

A convenient way of studying the friction of polymers is to slide a hard steel ball over a flat smooth strip of the polymer. The first thing we note is that a gentle groove is formed in the surface of the polymer. From the width of the groove we can make a rough estimate of the area of contact A between the ball and the polymer during sliding. Of course this does not distinguish between the apparent and real area but since the polymer is soft and smooth the asperities flow easily, so

TABLE II

FRICTION OF POLYMERS SLIDING ON THEMSELVES
Sphere on flat. Load 1 lb. weight. Sliding speed 1 mm./sec.

Polymer	Usual name	Chemical formula	Coefficient of friction μ_k
Polyvinyl chloride	PVC	$[-CH_2-CHCl-]_n$	0.4–0.5
Polystyrene	Polystyrene	$\begin{bmatrix} C_6H_5 \\ \vert \\ CH_2-CH- \end{bmatrix}_n$	0.4–0.5
Polymethyl methacrylate	Perspex, Plexiglas	$\begin{bmatrix} CH_3 \\ \vert \\ -CH_2-C- \\ \vert \\ COOCH_3 \end{bmatrix}_n$	0.4–0.5
Nylon		$[-CO-(CH_2)_4-CO-NH-(CH_2)_6-NH-]_n$	0.3
Polyethylene	Polythene	$[-CH_2-CH_2]_n$	0.6–0.8
Polytetrafluoroethylene	PTFE (Fluon, Teflon)	$[-CF_2-CF_2]_n$	0.05–0.1

that the difference cannot be large. The second thing we measure is the frictional force F. If we again write F = As we can calculate s, the specific shear strength of the interface. It turns out that s is very near to the bulk shear strength of the polymer.

This conclusion suggests that the adhesion at the interface is so strong that shearing actually occurs in the polymer itself and the strength properties of the polymer determine the friction. If we examine the steel ball we will observe a thin film of polymer attached to the surface of the slider at that region where sliding contact occurred. Further, we shall occasionally find a minute piece of steel transferred and attached to the polymer strip. It is not easy to explain this: probably, since steels are not very homogeneous the polymer is able to pluck

Friction of Non-Metals 79

out a portion of metal (or oxide) from a weak region in the steel surface. Whatever the explanation the observation is important, for it indicates that a relatively soft plastic can produce some wear of an apparently hard metal.

Sometimes polymers contain small quantities of materials called "plasticizers," which are introduced to enable the polymer to flow more easily. Sometimes other additives may be included for other purposes. In some circumstances these materials may diffuse to the surface and greatly reduce the adhesion. The friction may then be very much reduced. This behavior is often observed with many commercial forms of polythene which contain a plasticizer known as oleamide. This observation suggests interesting possibilities of producing "self-lubricating" polymers. If we add a small amount of suitable additive to the polymer before it is fabricated in bulk it may retain both the bulk properties of the polymer and the surface properties of "lubricated polymer." As the surface film is worn away, further material may diffuse to the surface to replenish the lubricant. Experiments along these lines in Britain and America indicate that this idea has practical possibilities.

Effect of Load and Geometry on the Friction of Polymers

One major difference between the frictional behavior of polymers and metals is the effect of load and geometry. By geometry we mean the shape of the surfaces, whether they are flat or curved and, if they are curved, how sharp the curvature is. With metals the deformation at the contact region is plastic and the area of true contact is determined only by the load and the yield pressure and not by the shape of the surface. Consequently with metals the friction does not depend on the geometry of the surfaces. With polymers it is otherwise. These materials deform visco-elastically: the deformation depends on the load W, the geometry and the time of loading. For a fixed time of loading and for a fixed geometry—a sphere on a flat surface (or a cylinder pressing on another cylinder at right angles to it)—the area of true contact is not proportional to the load as it is with metals, it is proportional to W^n where n is less than 1 and usually is more nearly ¾. Consequently the frictional force F varies as $W^{3/4}$. The consequence of this is seen by the following example, illustrated in Figure 31. Suppose when the load is 1 lb. the frictional force is 0.7 lb. (point A). If we increase the load 16-fold the friction will increase by the ¾ power of 16, which is 8. This means that for a load of 16 lbs. the frictional force has increased to 5.6

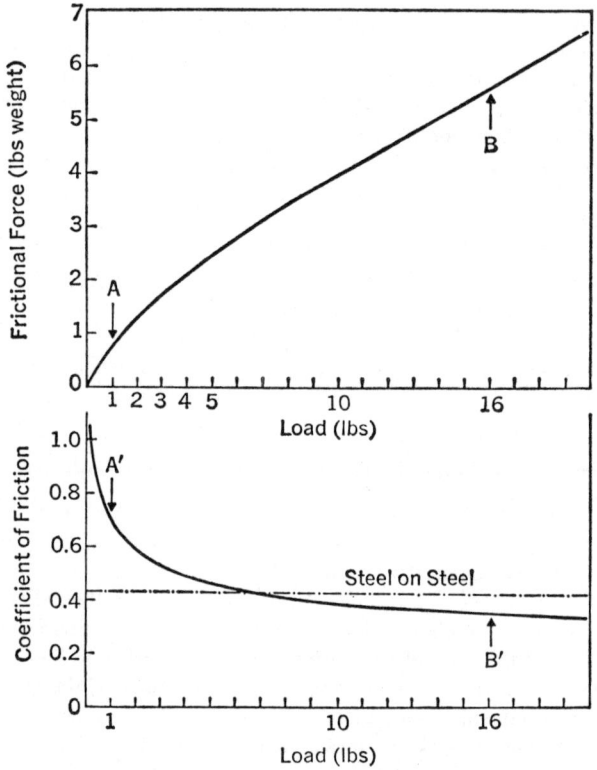

Figure 31 *Friction of a hard sphere sliding over a clean flat surface of Plexiglas. (a) The frictional force is not proportional to the load W but increases more nearly as W^n where n is about ¾. (b) As a result the coefficient of friction decreases as the load increases. The broken line represents results for steel on steel for which the coefficient of friction is almost independent of load.*

lbs. (point B). If a 16-fold increase in load produces an 8-fold increase in frictional force the *coefficient* of friction is *reduced* by a factor of 2 (points A' and B'). This is a general feature with polymers: the coefficient of friction gets smaller at higher loads.

If the surfaces are flat so that contact occurs at a large number of asperities the area of contact and the friction are more nearly directly proportional to the load. Consequently in most engineering applications the effect we have been discussing above is not of major importance. It is, however, of great importance in the textile industry, where fibers slide over one another or over the "guides" used in textile machinery. Here the loads are extremely small and, for the same reasons as those developed in the preceding paragraph, this can lead to a large and significant increase in friction. Values of μ greater than 1 can easily occur. For this reason the values quoted in Table II must be understood to apply to a specific load, geometry and sliding speed.

The Friction of Teflon or Fluon

This material, known chemically as polytetrafluoroethylene, or PTFE, has remarkable frictional properties. In structure it is similar to polyethylene except that all the hydrogen atoms are replaced by fluorine. It is very difficult to stick other solids on to PTFE since the glue will not "hold." This poor adhesion of PTFE to other materials is, of course, widely exploited in non-stick saucepans. The poor adhesion has indeed provided in the past an explanation for its very low friction. Recently, however, some researches in Britain have shown that under special conditions it is possible to obtain good adhesion between a piece of PTFE sliding over a clean glass surface, even though the friction remains very low, $\mu \cong 0.05$. Marked transfer of PTFE to the glass occurs. A little contamination reduces the friction to about $\mu \cong 0.03$ but reduces the interfacial adhesion so markedly that no detectable transfer occurs. We may conclude that it is easy to get poor adhesion, little transfer and low friction with PTFE. However, even if adhesion is good the friction is still low. The reasons for this are not clear; probably the behavior is associated with the crystal structure and relative "smoothness" of the PTFE molecule.

Uses of PTFE

PTFE has found many practical applications, particularly if loads and speeds are not too high. Two special uses may be mentioned. First, it has proved very successful as a surface coating of skis in place of the conventional waxes, since it gives a low coefficient of friction on both dry and wet snow and on ice; that is, its friction characteristic is fairly constant whatever the nature of the snow or ice it slides on. A second and more important technological use is in bearings. Unfortunately PTFE in bulk is not very suitable as a bearing since it is not strong enough mechanically and it is a poor conductor of heat. Furthermore it has a high coefficient of thermal expansion, so that, if it were used as a bearing material, it would heat up, expand and stick. These difficulties are almost entirely overcome by incorporating it into the surface of a porous metal such as sintered bronze. The resulting material has the mechanical and thermal properties of the metal and the surface properties of PTFE. These bearings are being more and more widely used and operate surprisingly well as dry bearings, that is, without the application of any oil or other liquid lubricants. In Table III we compare the friction of this type of bearing material with that of the more conventional metallic bearing materials.

TABLE III

COEFFICIENT OF FRICTION OF STEEL SLIDING ON VARIOUS BEARING MATERIALS (UNLUBRICATED) AT ROOM TEMPERATURE

Bearing Material	Coefficient of Friction μ_k
Copper-lead	0.2
White metal (lead or tin base)	0.5
Bronze	0.4
Bronze impregnated with PTFE	0.05 to 0.1

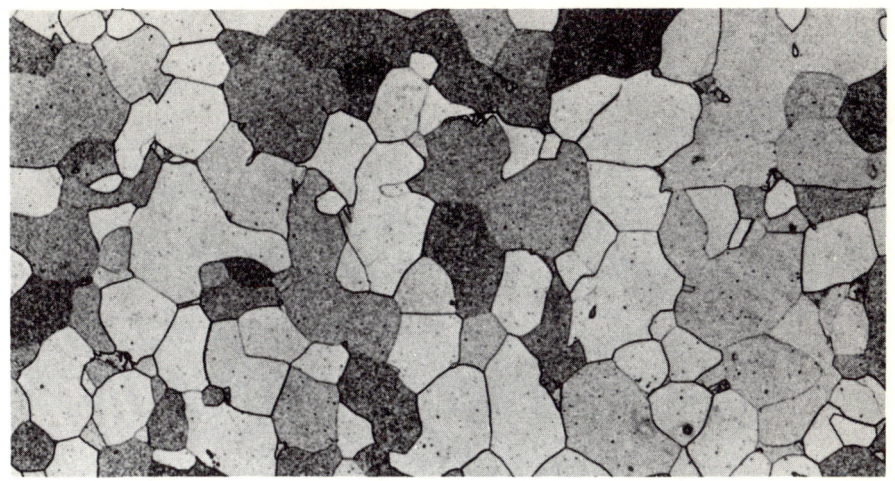

(a) A polished surface of copper has been "etched" with a suitable chemical which attacks the crystals to varying degrees depending on their orientation. It is seen that the material is polycrystalline: each crystal is about one thousandth of an inch across.

(b) A model of small spherical bubbles, all of equal size, floating on a water surface. They pack together in a regular array and so resemble the atoms in a crystal. The figure shows how various crystals or grains fit together. Triangles have been drawn joining rows of atoms in each crystal: these show the orientation of the individual crystals. It is seen that the crystals are differently oriented. The atomic arrangement in the grain boundaries, that is, where the individual grains fit together, is very irregular. (Based on the famous paper by Sir Lawrence Bragg and Professor John F. Nye on a bubble model of crystal structure.)

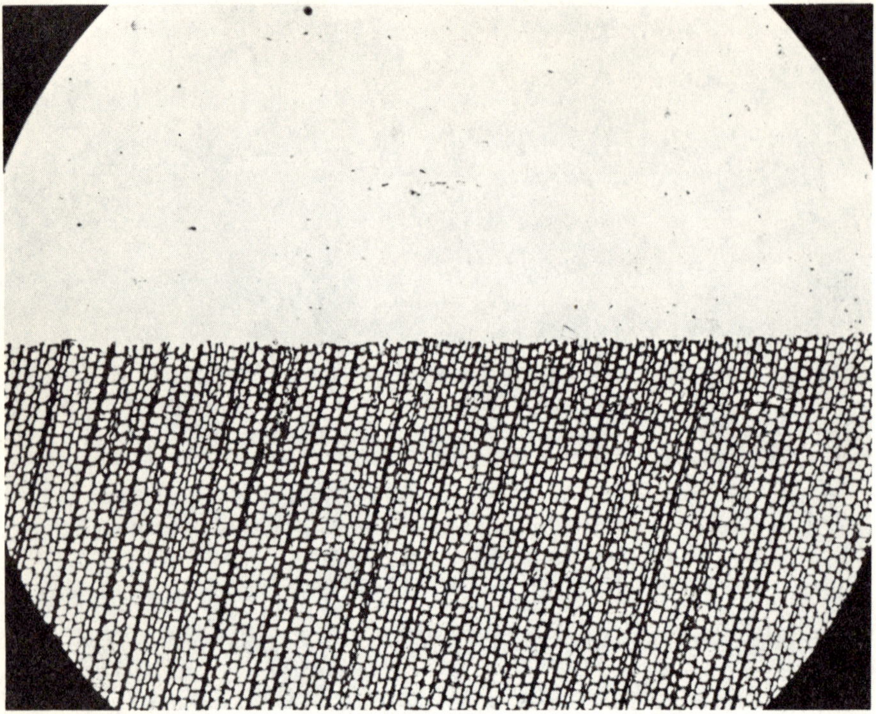

(a) Section through a specimen of balsa wood showing the arrangement of fibers. Each fiber is about one thousandth of an inch across.

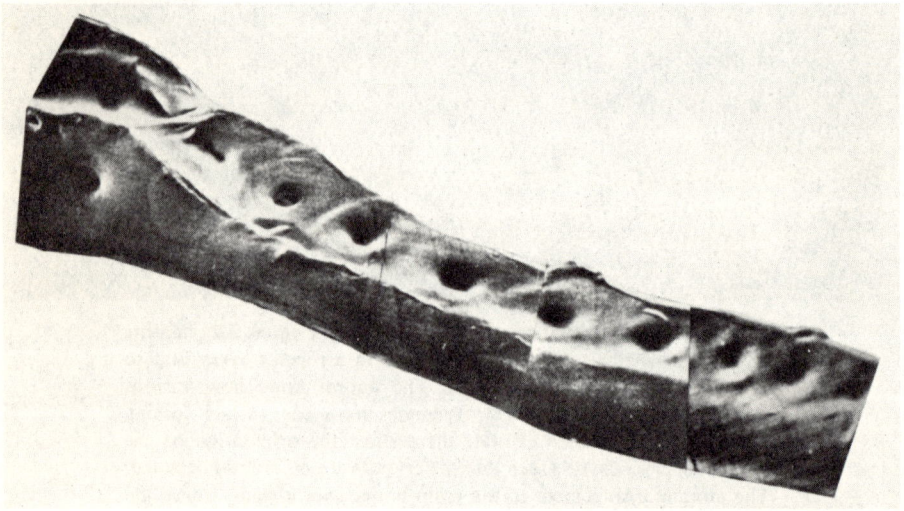

(b) A single wood fiber. The dark circles are little pits which act as valves to regulate water flow between neighboring fibers.

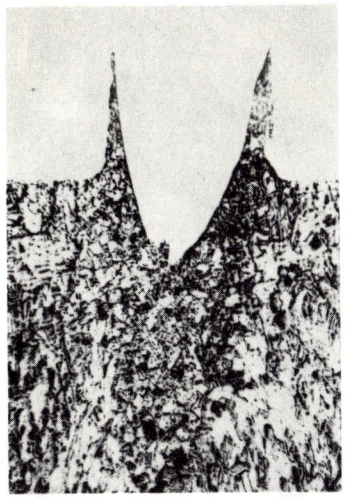

(a) Oblique section of pin scratch in copper showing raised edge and deformation below the surface. Vertical magnification 2,000 times.

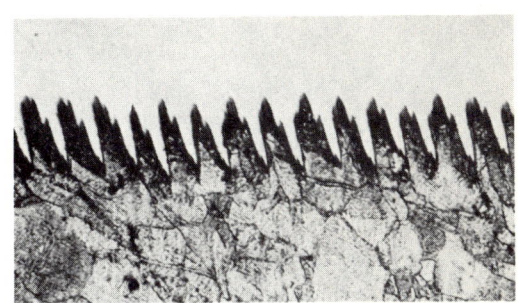

(b) Finely turned copper surface. Surface roughnesses are about 200 microinches. Vertical magnification 2,000 times.

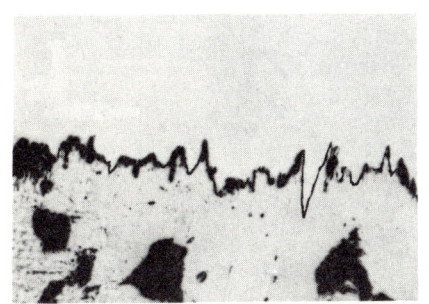

(c) Finely abraded steel surface. Irregularities about 20 microinches. Vertical magnification 1,200.

(d) Finely polished steel surface. Irregularities about 4 microinches. Vertical magnification 1,200.

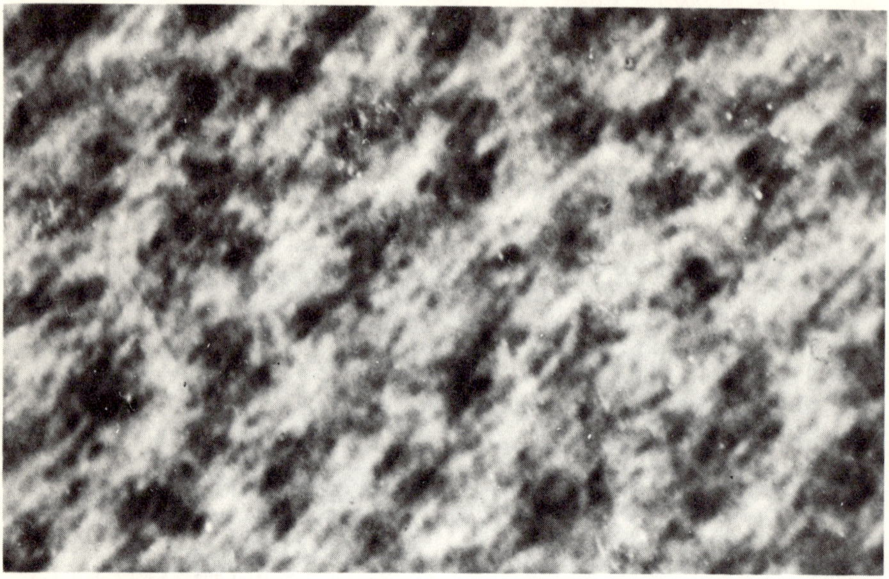

Transmission Electron Microscopy (Using Replicas)

(a) Electrolytically polished aluminium. Magnification 100,000 times. Shadows at edge of hills show that irregularities are between 100 and 1,000 Å high.

(b) Electrolytically plated gold (by kind permission of Dr. J. B. P. Williamson). Magnification 4,000 times. Hills are about 1,000 Å high.

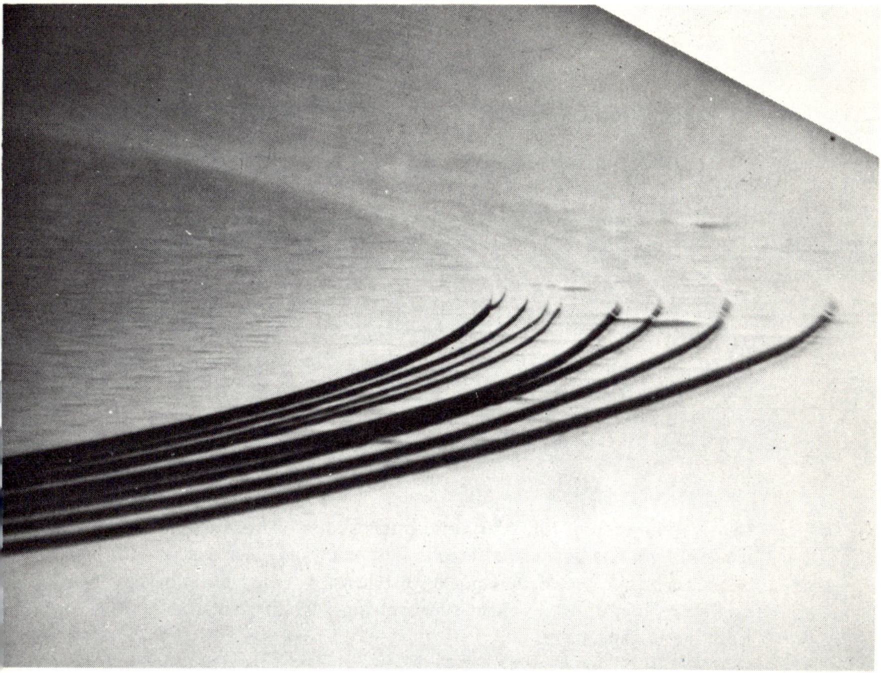

(a) Cleavage surface of mica at glancing incidence. The cleavage steps are about 1,000 Å high.

(b) Friction track formed in copper surface by a curved slider. Load 20 gm. Horizontal magnification 3,000.

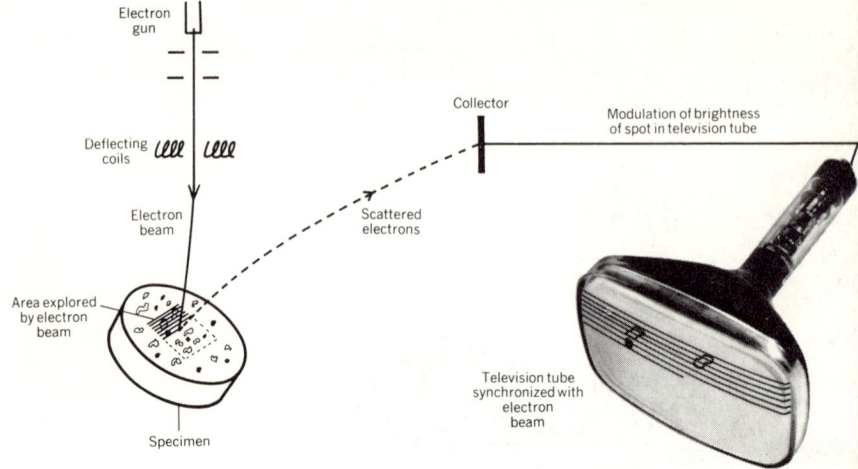

(a) Arrangement of scanning microscope. The specimen is scanned by a fine beam of electrons. The scattered electrons "modulate" the brightness of the spot on a television tube; the television spot scans the screen in exact synchronism with the electron beam scanning the specimen.

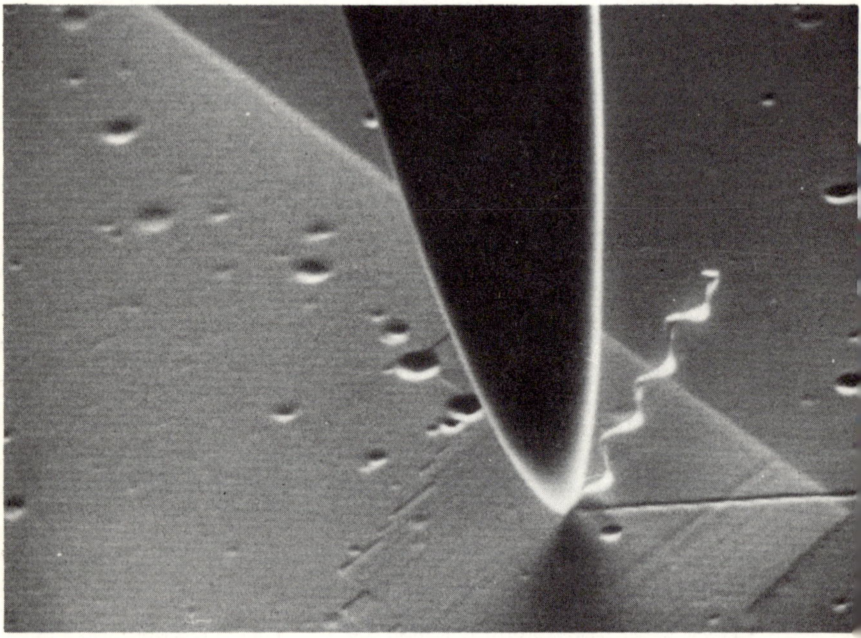

(b) Scanning electron microscope picture showing a pointed slider traversing a copper specimen. The slider produces a fine shaving and a groove. Load 0.1 gm. Horizontal magnification 3,000.

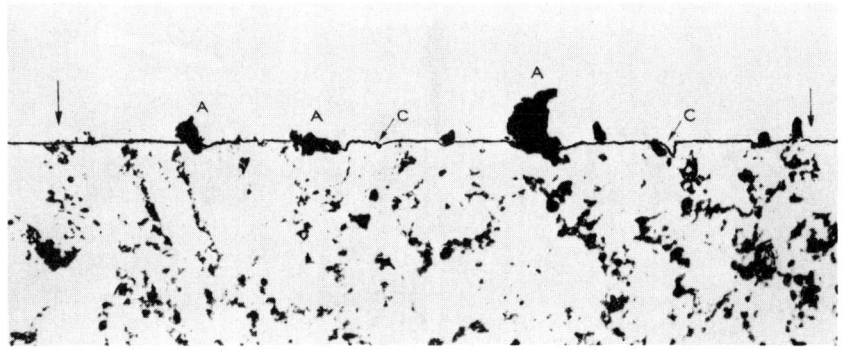

(a) Taper section of track formed on steel when a curved copper slider slides over it (unlubricated). Large fragments of copper (A) are left adhering to the steel surface: occasionally small fragments of steel have been plucked out of the steel surface (C). Horizontal magnification 200.

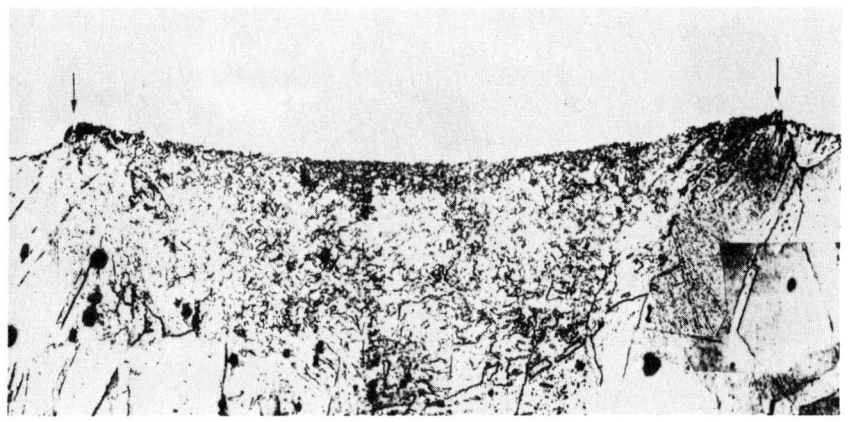

(b) Taper section of track formed on a copper surface when traversed by a hard steel slider (unlubricated). There is plowing of the surface and heavy subsurface deformation. Horizontal magnification 400.

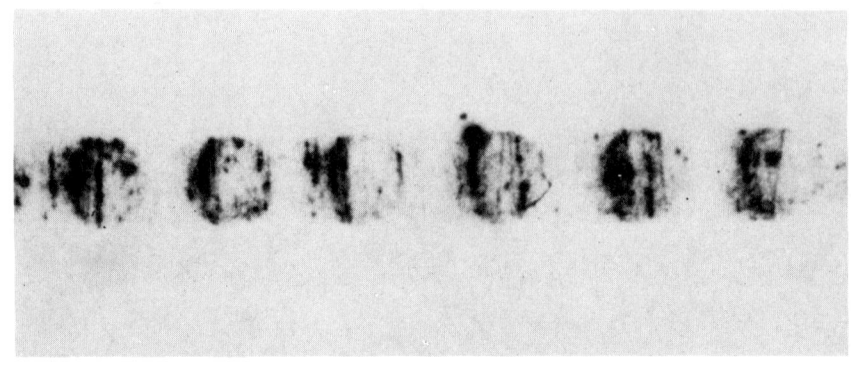

(a) Autoradiograph of track formed on steel when lead slides over it (unlubricated).

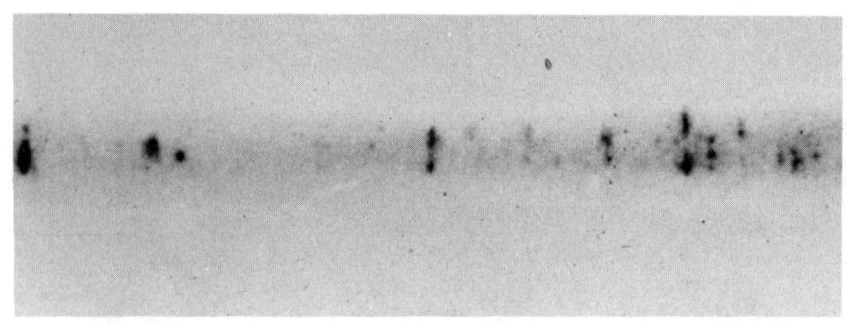

(b) Autoradiograph of track formed by lead sliding on a lubricated steel surface. Metallic pickup is reduced but not eliminated.

Friction of Rubber

Rubber is an extreme example of a material that deforms elastically. However much we distort it, provided we do not actually tear or cut it, it will return to its original shape when the deforming force is removed. Contrast this with polymers such as polyethylene or nylon which deform elastically and also flow viscously; and with metals which deform elastically over a very limited range and then flow plastically. It is not surprising that the friction of rubber often shows an appreciable deviation from Amontons' laws. The friction depends markedly on load and on the geometry of the surfaces.

If we press a hard sphere on to a flat rubber surface the area of contact is proportional to $W^{2/3}$. This resembles the behavior of polymers. However, rubber is a truly elastic solid and the power of W is accurately $2/3$. By contrast polymers are visco-elastic solids and the power of W depends on the material: for most polymers it is nearer $3/4$ (0.75) or $4/5$ (0.8) than $2/3$ (0.67). If now we slide a hard sphere over a smooth flat rubber surface the area of contact and hence the frictional force is proportional to $W^{2/3}$ (see Figure 32). There may of course be a contribution from the grooving or plowing of the rubber surface by the hard sphere but for unlubricated surfaces this is generally small and can be neglected.* If the sliding surfaces are flat so that they touch over a large number of contact regions the area of contact and the friction are more nearly directly proportional to the load as we noted in our discussion of the friction of polymers. Over a wide range of practical conditions the friction of various types of rubber against a hard surface lies between $\mu = 1$ to $\mu = 4$.

The deformation properties of rubber depend on the rate of deformation and on the temperature in a well-defined way and it is found that the sliding friction reflects these properties. We shall discuss this in greater detail in a subsequent

* By contrast, when lubrication is really effective the major part of the friction may arise precisely from this plowing or deformation mechanism as we shall see in Chapter X.

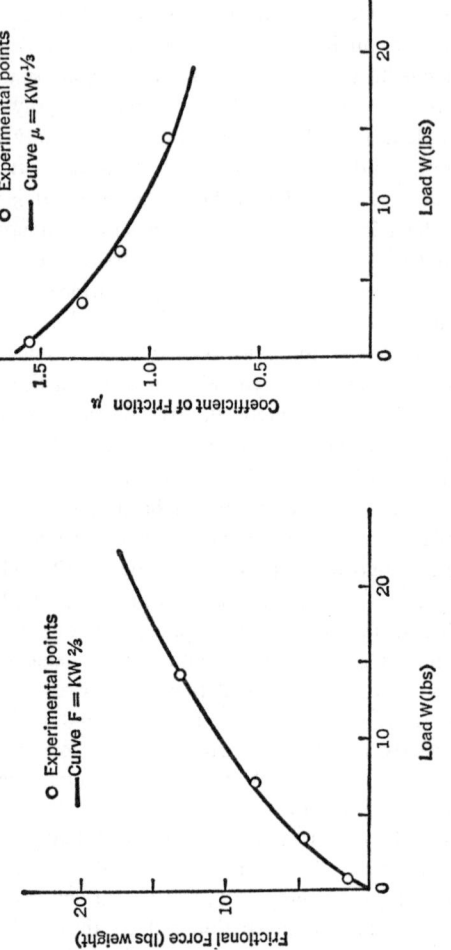

Figure 32 Friction of a hard sphere sliding over a clean flat rubber surface. (a) For clean surfaces the friction is dominated by the adhesion term. The area of contact, for elastic deformation, increases as $W^{2/3}$, so that the frictional force F varies in this way with the load W. (b) The coefficient of friction $\mu = F/W$ varies as $W^{-1/3}$, that is, it decreases as the load increases. Note the relatively large values of μ as compared with the behavior of lubricated rubber surfaces. See for example Figure 58.

chapter when we consider the frictional behavior of automobile tires.

Friction of Mica

Mica is an interesting material. It has a single extremely well-defined cleavage plane. As a result we can cleave sheets of mica which do not have any cleavage steps on them, that is to say, they may be *molecularly* smooth over areas of several square centimeters. When such surfaces are placed together they give molecular contact over the whole of the area where the two sheets overlap. Pressing the sheets together will not increase the true area of contact. As one would expect, the friction does not increase: indeed, the friction is almost independent of load. On the other hand, increasing the area of overlap increases the friction in the same proportion. We may say that whereas with polymers and rubber the frictional behavior *deviates* in a fairly well-defined way from the laws of Amontons, with smooth mica sheets the frictional behavior bears no resemblance to Amontons' laws. The friction is high, does not depend much on the load and is proportional to the size of the surfaces. We observe a similar trend in the frictional behavior of very thin polymer foils.

Friction of Wood

Wood was one of the materials used by Amontons in his classical experiments on friction in 1699, but very little was known of its chemical composition or structure. We now know that both its composition and structure are very complicated. As mentioned in Chapter III, wood is composed of fibers consisting largely of cellulose: these are packed side by side with occasional cross connections, the fibers being stuck together by lignin and other gluelike materials. The friction of wood on wood or metal on wood gives very erratic results because of the presence of small amounts of wood fats. If these are removed the friction is much more reproducible. The deformation properties of wood are like those of polymers and the friction coefficient depends to some extent on load and geometry. However, over a fairly wide range of conditions for wood on wood and for metals on wood the friction for clean surfaces is about $\mu = 0.5$. If a hard slider is used, an appreciable part of the friction may be due to grooving or plowing.

There is appreciable adhesion between wood and other surfaces due to the presence of "hydroxyl" (OH) groups in the cellulose. This accounts for the greater part of the friction. If the wood is wet, the friction falls by about 20 per cent. We observe a much more marked effect in the sliding of PTFE on wood. With dry wood the coefficient of friction is about $\mu = 0.25$; on wet wood it falls to $\mu = 0.1$. This is because of the weak adhesion between PTFE and the surface water-film.

The friction of lignum vitae is of particular interest. It is an extremely hard dense wood with a large content of natural waxes. As long ago as 1717, James Harrison, the clockmaker who pioneered the first timepiece suitable and accurate enough for navigation at sea, used lignum vitae in roller pinions because of its low friction. "These Rolls of Wood," he

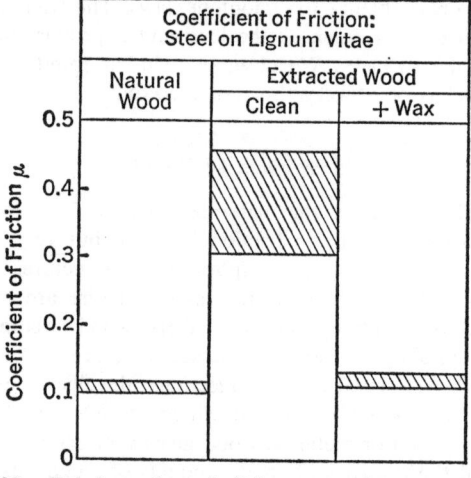

Figure 33 *Friction of steel sliding on a flat strip of lignum vitae showing results obtained with natural wood, with wood from which the natural waxes have been extracted and with such wood when the extracted waxes are reapplied to the surface.*

wrote, "move so freely as never to need Oyl." Even today it is still used for the propeller bearings of ocean liners. The friction of wood or metal on natural lignum vitae is very low, of the order $\mu = 0.1$. This is due to the waxes, which are very hard to remove. If they are extracted with suitable solvents the friction rises to about $\mu = 0.5$, i.e., the value of ordinary woods. If the extracted wax is spread over the wood the friction falls to its original low value (see Figure 33). With natural lignum vitae the friction is not only low, it remains low with repeated traversals of the surface. This is because there is a three-dimensional supply of wax within the wood that can be squeezed out and supplied to the rubbing surface. It is, in effect, a self-lubricating polymer. Partly because of its high wax content it is very little affected by water. For this reason and because of its toughness and low friction it is well suited as a marine bearing material.

Brittle Solids

Our picture of the sliding process includes two main ideas. The first is the elastic or plastic or visco-elastic deformation of the surfaces at the points of contact. The second is the strong adhesion at the interface which results in the formation of friction junctions that have to be sheared if sliding is to occur. At first sight it seems that with brittle solids neither of these ideas would be particularly appropriate. One expects the surface to crack and fragment at the regions of contact, giving a completely irreproducible area of contact. Again, one hardly envisages adhesion between hard brittle solids.

Experiments however, show that intuition here is quite unreliable. First, the friction of brittle solids is fairly reproducible and of the same order of magnitude as for metals. Second, at the regions of contact the high local compressive stresses inhibit brittle fracture. The material, which in the absence of these stresses may be very brittle, flows plastically at the contact regions, giving an area of contact proportional to the load, as with metals. Further, strong local adhesion occurs. During sliding the surfaces may indeed crack and fragment—but this occurs mainly after the sliding process has taken place:

the intrinsic sliding mechanism is almost exactly the same as for metals and other ductile solids.

There is, however, one very marked difference. With thoroughly clean metals (see Chapter VII) gross seizure may occur. This is because strong adhesion combined with high ductility leads to a growth in the area of true contact as sliding takes place and this in turn leads to enormous coefficients of friction. With brittle solids plastic flow around the contact region is more difficult and, because of limited ductility, junction growth on an appreciable scale is not possible. The friction rarely exceeds $\mu \cong 1$.

The Friction of Diamond and Other Hard Solids

The friction of solids on diamond is generally small. However, it depends on the load and geometry because elastic rather than plastic deformation determines the area of contact. As with rubber and polymers the coefficient of friction decreases as the load is increased.

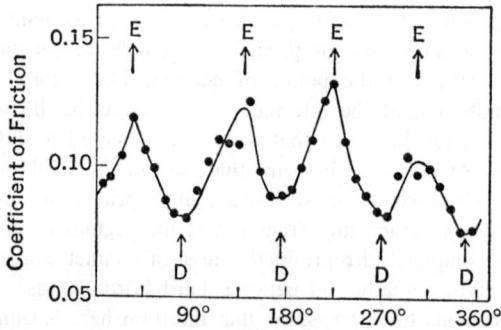

Figure 34 *Friction of a pointed diamond slider over the cube face of a natural diamond. The friction varies with the direction of sliding. It is highest when sliding takes place along a direction parallel to the cube edge E, E, E, E, and lowest when the direction is parallel to the cube diagonal D, D, D, D. The low friction direction is the hard direction, that is, it is the direction along which it is very difficult to abrade away the diamond surface. The high friction direction is the soft direction and is the direction chosen for easy removal of material during abrasion or polishing.*

Friction of Non-Metals

For diamond sliding on diamond at moderate loads the coefficient of friction in air is about $\mu \cong 0.1$. The friction, however, is markedly dependent on the crystal face and the direction of sliding (see Figure 34). These are associated with directions where the diamond is known to be hard or soft. In the soft directions the friction is high ($\mu \cong 0.15$); in the hard directions, low ($\mu \cong 0.05$). It is almost certain that this is due to the penetration of the diamond in the soft direction: this gives a larger area of contact and an increase in the amount of plowing.

The reason for the low coefficient of friction of diamond and other jewels such as garnet or sapphire is not fully understood. It is probably due to the presence of adsorbed surface films of oxygen or water vapor only one or two molecules thick. If these are removed, for example, by carrying out experiments in a high vacuum, the friction can rise to very high values (see Chapter VII) and appreciable wear and surface fragmentation can occur.

Friction of Ice and Snow

The low friction of ice is due to the presence of a thin lubricating water film on the surface. Experiments that show this are given in Figure 35. In these experiments miniature skis covered with Plexiglas, ski wax and PTFE were slid on snow at $-10°C$, on dry sand and on snow near $0°C$. If the friction is measured on snow at $-10°C$ under conditions where water film formation is difficult, the friction is comparable with that observed on dry sand. For Perspex skis at this temperature $\mu \cong 0.4$, for surfaces covered with ski wax or ski lacquer the friction is a little lower. Only with PTFE surfaces is the friction low, $\mu \cong 0.1$. On snow near $0°C$ where a water film is easily formed all skis show a low friction ($\mu \cong 0.05$).

What is the factor mainly responsible for the formation of the water film? At one time it was thought this was due to the high pressures at the points of contact which cause the ice to melt. We now know this is not the main cause. The main cause is the sliding process itself, which heats up the surface and causes melting in a very thin surface layer. Consequently high sliding speeds, which favor the generation of frictional heat-

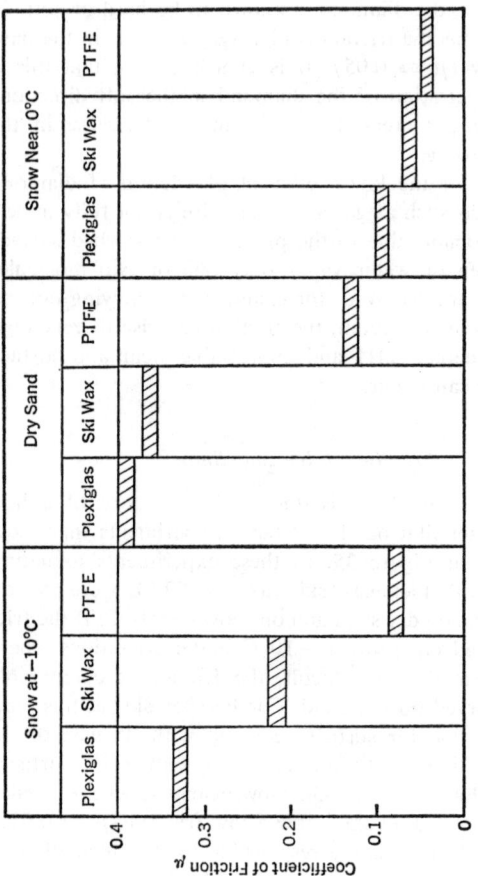

Figure 35 *Friction of miniature skis covered with Plexiglas (Perspex) ski wax and PTFE (a) on snow at $-10°C$; (b) on dry sand; (c) on snow near 0°C. On snow at $-10°C$ the friction for Plexiglas is almost as high as on dry sand. Only PTFE gives a low friction ($\mu \cong 0.1$). On snow near 0°C where a surface film of water is easily formed all skis show a low friction (μ lies between 0.06 and 0.1). The friction of PTFE is least dependent on the nature or condition of the other surface.*

Friction of Non-Metals

ing, tend to give a lower friction on snow and ice. Similarly, the friction for all materials sliding on ice or snow is small, very near 0°C. This is summarized in Figure 35 and shows that the friction of PTFE varies far less with surface conditions than other sliding materials.

Friction of Layerlike (Lamellar) Solids

There are a number of substances both natural and synthetic which have a layerlike or lamellar crystal structure. This is shown in Figure 36 for graphite and in Figure 37 for

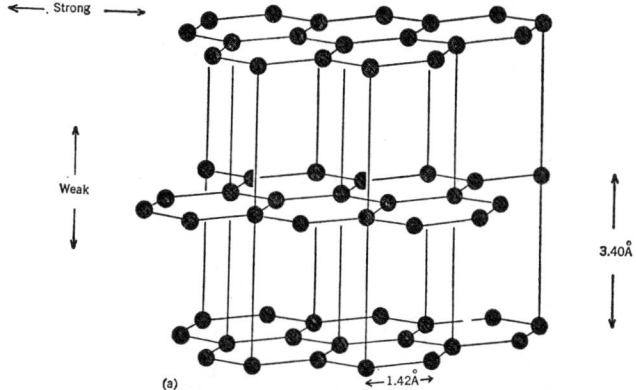

Figure 36 Structure of graphite.

molybdenum disulphide (MoS_2): they are composed of sheets or layers. The layers themselves are strong but the bonding between the layers is weak. Consequently these materials are strong in compression but weak in shear.

Graphite is carbon in a layerlike form: if slid on another clean surface its coefficient of friction over a wide range of conditions is about $\mu = 0.1$ to 0.15. As mentioned above, the graphite crystal is strong in compression across the layers but the layers slip easily over one another. This is favorable for producing a low friction. However, recent work shows that this explanation is an oversimplification. The low friction is partly due to the presence of very thin adsorbed layers of oxygen and water vapor on the edges of the platelets. If these

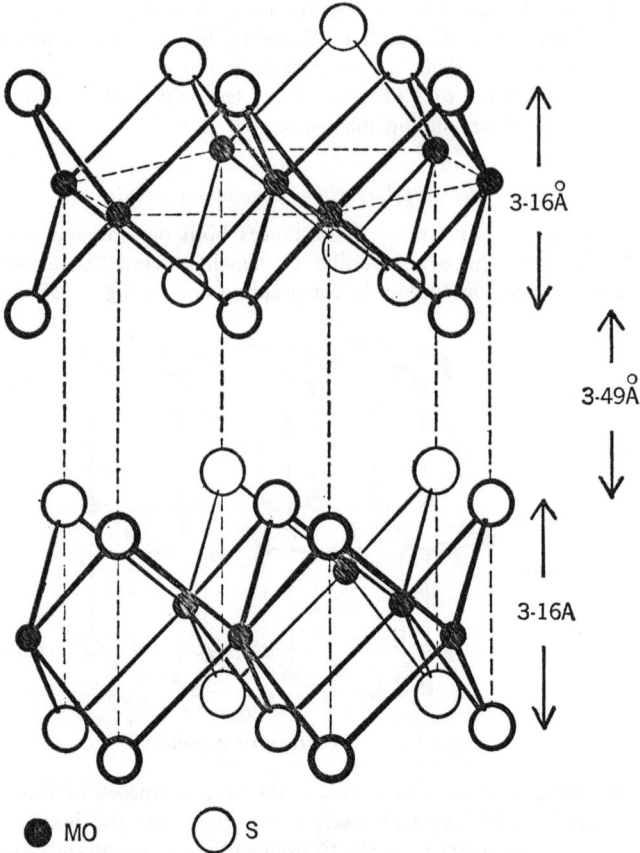

Figure 37 *Structure of MoS_2. Both graphite and MoS_2 are strong in compression but weak in shear.*

films are removed the friction rises ($\mu \cong 0.5$) and there is an enormous increase in the rate of wear. This result had an important repercussion on the performance of carbon brushes in electrical generators fitted to high-flying aircraft in World War II. In the rarefied atmosphere at a height of 30,000 feet any existing adsorbed films are worn away by the sliding process and do not have time to re-form. The life of a carbon

Friction of Non-Metals

brush in these circumstances is only a few minutes. The problem was finally solved by incorporating certain lubricating materials into the brushes during manufacture.

Molybdenum disulphide is another material which has a layerlike structure (see Figure 37). It has the advantage over graphite that its low friction ($\mu \cong 0.2$) does not depend on the presence of adsorbed films. (It cannot be used as brushes since it is electrically non-conducting). However, at very high temperatures graphite is superior. Above 800°C, MoS_2 decomposes in vacuum to leave metallic molybdenum and the friction rises to the high value characteristic of clean metals: with graphite the friction falls from $\mu = 0.5$ at room temperature to $\mu = 0.2$ at 2000°C. Of course in air the whole process is greatly complicated by oxidation.

And now, a cautionary note. Because graphite and MoS_2 are such effective solid lubricants it has generally been considered that all solids with a lamellar structure will have low-friction properties. This is by no means true. For example, titanium disulphide (TiS_2) has a lamellar structure very similar to that of MoS_2: yet it has poor anti-friction properties presumably because the bonding across the lamellae is relatively strong. Again, mica has a marked layerlike structure but is not suitable as a low-friction material. Here again the bonding across the layers is very strong. If the *chemical* composition of mica is suitably modified the bond strength can be greatly reduced and in that case a low friction solid is obtained: the resulting material is talc. We may conclude that lamellar solids are materials most likely to have low-friction properties but only if the inter-layer bonding is weak.

The lamellar solids most widely used for their low-friction properties are graphite and MoS_2. These materials are often applied to sliding surfaces to form friction-reducing coatings. The main difficulty is to "rub-them-in"; the particles stick to the surfaces by embedding themselves in. Another method is to bond them on to the surface with a suitable resin. Yet a further approach is to use a sintered porous metal (see Chapter V) and to incorporate the graphite or MoS_2 into the pores. An interesting innovation is to make one of the surfaces of sintered molybdenum and to treat this chemically so as to form MoS_2 *in situ*.

Composite low-friction surfaces of this type are finding increasing application under conditions where conventional lubricants are ineffective or undesirable. A similar approach has been used in the search for materials that will slide on one another successfully at very high temperatures.

VII FRICTION UNDER EXTREME CONDITIONS

> All things and everything whatsoever however thin it be which is interposed in the middle between objects that rub together lighten the difficulty of this friction.
> Leonardo da Vinci

Friction in High Vacuum: Surface Cleaning by Heating

Surfaces normally encountered in air are covered with thin films of oxides, adsorbed layers of oxygen and thin layers of water vapor. Contact between surfaces in air invariably traps these surface films between them; this weakens the adhesion at the interface.

Surface films are removed partially or completely in a number of ways. One way of doing this in air is to increase the load between the sliding bodies. Very heavy deformation at the regions of contact may then rupture the surface films. If the surface film is an oxide which has a high electrical resistance this effect may be demonstrated in a simple way by measuring the electrical resistance between the surfaces. For example, with copper sliding on copper at low loads the electrical resistance is high and the friction is of the order of $\mu = 0.4$; evidently sliding is occurring within the oxide film itself. If the load exceeds a critical value the electrical resistance across the surfaces suddenly drops and the friction rises to about $\mu = 1$ (see Figure 38). Clearly when the oxide film is broken down adhesion between the surfaces is much stronger than when it is present.

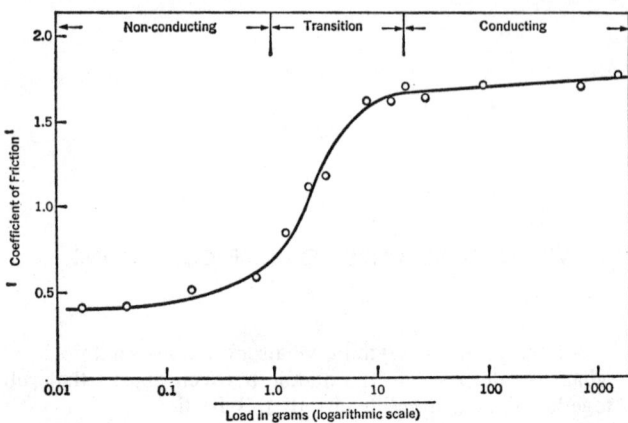

Figure 38 *Breakdown of oxide film on copper as a function of load when a hard steel ball slides over the surface. Below 1 gm. the friction is low ($\mu \cong 0.4$) and the surfaces are electrically insulating; sliding occurs within the oxide film on the copper surface. Above 100 gm. the friction is high ($\mu \simeq 2$) and the surfaces become electrically conducting. The oxide is penetrated giving metallic contact and increased interfacial adhesion.*

A far more effective way of removing surface films is by heating the bodies in very high vacuum. It is now possible to buy "off the shelf"—if there is enough money in the kitty— standard equipment which will produce a vacuum in which the air is one ten million millionth (1/10,000,000,000,000) of an atmosphere. If surfaces are heated to drive off or decompose surface films they will remain clean in such a vacuum for several hours. Results show that under these conditions the friction between most metals reaches enormous values of the order $\mu = 10$ or 50 or more. Gross seizure occurs: that is to say, both the friction and the surface damage are very large indeed. The reason is simple. When we first place the clean surfaces together the area of contact is determined by the plastic yielding of the tips of surface asperities as described in Chapter IV. At these regions the naked surfaces adhere extremely strongly. When we apply a tangential force

Friction Under Extreme Conditions

to start sliding the surfaces over one another we produce further flow around the regions of contact. Because the surfaces are so clean, wherever they touch they stick very strongly. As a result "junction growth" occurs before actual sliding, on an observable scale, takes place. The junction area can increase until the whole of the geometric area of contact becomes an enormous friction junction. The friction may then bear little relation to the original load: coefficients of 50 or more may be observed. At the same time the surfaces are heavily damaged.

Junction Growth and How to Overcome It

There are two factors which can reduce "junction growth." The first is contamination. A whiff of oxygen which forms a surface layer of oxide only a few molecules thick can reduce the friction from enormous values to the more usual values ($\mu = 0.5$ to 1) encountered with metals in air. The surface film prevents the surfaces from sticking together strongly and allows only a small amount of junction growth to occur. Clearly the presence of air is a very fortunate arrangement in normal running machinery. It not only keeps people alive, it keeps most sliding mechanisms viable. Readers of Henry James may be entertained by the following quotation from *The Portrait of a Lady*, Chapter 6; it has no scientific significance and yet it summarizes the tribological situation rather well. The phrase runs, ". . . in the thick mild air all friction dropped out of contact."

The majority of metals used in machinery readily form surface oxide films. If they did not they would be far more difficult to use even in the presence of lubricants. Metals such as gold and platinum (known as "noble" metals because they do not react readily with their surroundings as do the "base" metals—iron, copper, lead) do not form surface oxides. Their behavior in air resembles that of base metals examined in a high vacuum where the oxide films have been removed. Thus, even in air, their friction is very high and marked adhesion, wear and transfer of metal occur during sliding. Even if they were cheap and had the appropriate mechanical properties they would not be suitable for sliding mechanisms. Indeed the lubrication of gold surfaces used in delicate electrical contacts raises very special problems which have only recently been solved.

The second method of restricting "junction growth" is to use materials of limited ductility. These materials after a small amount of junction growth will fracture rather than flow further. Even when the surfaces are thoroughly clean they may therefore give a lower coefficient of friction ($\mu = 1$ to 2). There is, however, the danger that they may undergo serious cracking and fragmentation.

Friction in High Vacuum: Surface Cleaning by Rubbing

The cleaning of surfaces by heating to a very high temperature in vacuum may, of course, produce changes in the structure of the solid itself. For example, with diamond the strongly adsorbed surface films normally present can only be removed by making the diamond so hot that the surface layers turn into graphite. As a result the friction does not rise

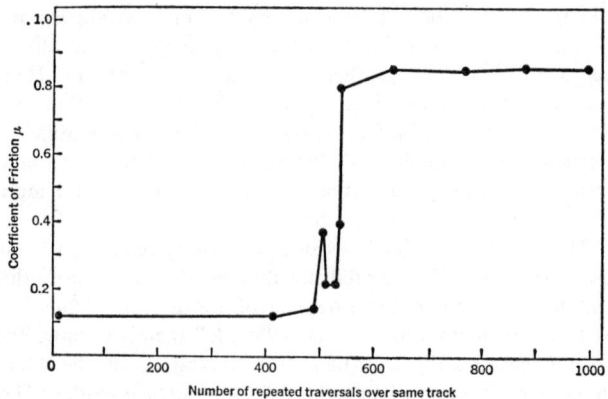

Figure 39 *Friction of a pointed diamond slider traversing a diamond surface in a very high vacuum. After repeated traversals of the same track the protective films naturally present on the diamond are worn away and have not sufficient time to reform. The friction rises by a factor of about 10 and there is heavy surface fragmentation of the diamond.*

Friction Under Extreme Conditions

above about $\mu = 0.4$. One way of overcoming this difficulty is to slide the surfaces repeatedly over one another in a very high vacuum so that the surface layers are gradually worn away and cannot reform. If this is done with diamond the friction, after a few hundred traversals, reaches values of the order of $\mu = 1$ (see Figure 39). Further, the increased adhesion at the interface tears lumps of diamond out of the surface and rapid fragmentation and wear occur. Clearly there is no future for diamond bearings in high vacuum applications.

Friction in High Vacuum and at Very Low Temperatures

Systems which operate in outer space have two difficulties to contend with. First, the almost complete absence of any atmosphere. Second, the temperatures are extremely low, of the order of $-270°C$, almost the absolute zero. Experiments under these conditions show that at these low temperatures many metals lose their ordinary ductility; they deform plastically much less readily. What is the implication of this? It means that when we try and slide one body over the other, the junction growth mechanism which we described above is much more restricted. Some junction growth does occur but on a reduced scale. Under these circumstances the friction may be much less than at room temperature. For example, even with clean gold surfaces in high vacuum the friction may not exceed $\mu = 1.5$.

These results suggest that one may be able to find solids which will show a relatively low coefficient of friction if they are intrinsically lacking in ductility. They must, however, not be brittle, for then they will fragment as diamond does. Experiments both in the United States and in Britain indicate that this may provide a fruitful approach. Certain metal alloys, particularly those possessing "hexagonal" structure (their structure resembles that of zinc, shown in Chapter III, Figure 15c), often show this property. It would seem that these materials offer certain advantages as frictional surfaces for operation without lubricants under "outer space" conditions.

Friction at Very High Speeds: Frictional Heating

The energy lost during sliding appears mainly in the form of heat. This has long been known and from ancient times the rubbing together of two bad thermal conductors such as two pieces of wood has been used as a means of kindling a fire. For example, in the writings of Tsou Yen, who lived around 300 B.C. in the Chinese state of Chi, we find the following comment: "In Spring, fire should be kindled by twirling the fire drill in elm and willow-wood. In Summer the wood of the jujube-tree and the apricot tree should be used: in Autumn the oak and the *yu*-tree. In Winter one must make use of the wood of the *huai*-tree and the *than*-tree." Presumably these different combinations of wood are meant to take into account the friction and humidity of the various species at different times of the year.

With good thermal conductors such as metals frictional heat can leak away more readily so that the overall heating will be considerably less. However, we must remember that the real area of contact is extremely small and it is precisely at these points that the frictional heat is liberated. Consequently the temperature at the points of real contact may still be very high, even though the bulk temperature rise may be small.

How are we to measure this temperature? We cannot use ordinary thermometers since they can never approach close enough to the rubbing interface. One very ingenious and useful method is to make use of the thermocouple effect. If, say, a piece of lead is joined to a piece of steel and the junction heated, a small voltage will develop at the junction which is a measure of the temperature. For example, a temperature rise of 100°C will produce a thermal voltage of about 4 millivolts. This can easily be measured with a sensitive galvanometer or a cathode ray oscilloscope.

This at once suggests that if we slide a piece of lead over a steel surface and measure the thermoelectric voltage we shall acquire a very good idea of the temperature at the interface (see Figure 40). Experiments show that this is so. For example, Figure 41 shows results obtained for various metals

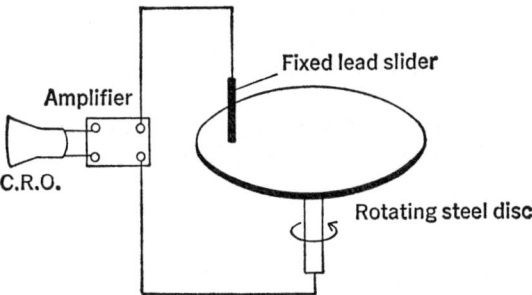

Figure 40 *By sliding lead on steel we may use the surfaces as their own thermocouple and so measure on a cathode ray oscilloscope (CRO) the thermoelectric voltage generated during sliding. This is a measure of the temperature developed at the points of real contact.*

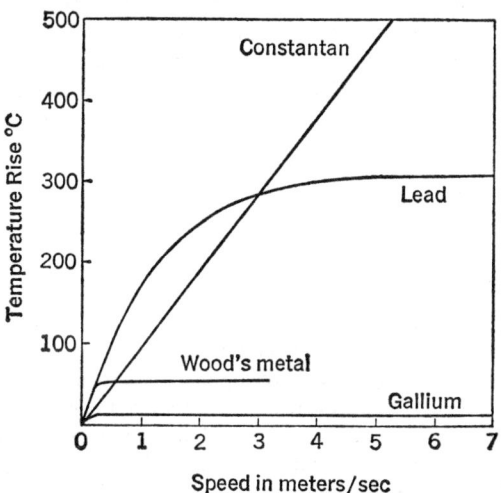

Figure 41 *Temperature rise against sliding speed for a number of metals sliding on steel using the thermocouple method shown in Figure 40. The temperature does not exceed the melting point of the metal. Melting point of gallium, 32°C; of Wood's metal, 72°C; of lead, 327°C; of constantan, 1290°C.*

sliding on steel: we see that the temperature increases with speed of sliding but does not generally rise above the melting point of the metal used. We also see that at quite modest sliding speeds (5 meters/sec. = 15 m.p.h.) surface temperatures of the order of 500°C may occur at the contact spots. Overall, the bodies may appear quite cool.

Polishing

These high temperatures at the regions of real contact must have a profound effect on the frictional process itself. There are three examples of this in everyday affairs. First, in converting a rough surface into a smooth polished surface. In the initial stages the surfaces are cut or abraded by the polishing powder to produce a finer and finer surface finish. At the final stages, however, local flow of the material occurs to produce the characteristic polish layer. In general, a polishing powder must have a higher melting point (or softening point) than the surface being polished. This shows that, broadly speaking, polishing is due to frictional heating, which produces a softened or molten layer which is smeared over the surface.

Machining

The second example is in the cutting of metals as in a lathe or milling machine. The tool removes a piece of metal which slides over the tool surface in the form of a shaving or chip. Here extremely high local temperatures can be reached particularly in the high-speed machining of hard materials. Steel cutting tools will overheat, lose their strength and blunt very rapidly. In such circumstances tool tips of materials possessing much greater strength at elevated temperatures must be used. These include diamond, tungsten carbide and titanium carbide. The behavior of tungsten carbide is of particular interest. In cutting non-ferrous materials it behaves extremely well even at very high sliding temperatures. With ferrous materials (e.g., steels) its performance can be very poor. The reason is that if the temperature of the rubbing interface exceeds 1300°C the tungsten carbide actually dissolves into

Friction Under Extreme Conditions

the work piece since the solubility of tungsten carbide in iron at these temperatures is appreciable. Under similar conditions titanium carbide does not dissolve in ferrous materials or does so at a very much lower rate. Consequently in machining ferrous materials at very high speeds with carbide tools the tool wear is greatly reduced if titanium carbide is used or incorporated as a component in the tool material.

Skiing on Ice and Metals

The third example is in the sliding on ice or snow as discussed in a previous chapter. The frictional process produces local melting and the thin film of water formed between the sliding surfaces is responsible for the very low friction observed. If the ice is too cold this does not readily occur and the friction remains high. Skiing is easy only if the ice or snow is not colder than a few degrees below zero.

It is natural to ask if such a process can also be observed with metals. The answer is yes—if the sliding speed is high enough. Figure 42 shows friction results obtained for a steel ball sliding on copper in a modest vacuum. At ordinary engineering sliding speeds, say, 50 m.p.h., the coefficient of friction is about $\mu = 3$. As the speed is increased the friction falls until at a speed of 2,000 m.p.h. the friction is less than $\mu = 0.2$. Examination of the surface shows that local melting has occurred and sliding at these speeds takes place in a very thin molten film formed by the sliding process itself. With metals possessing a lower melting point, such as bismuth, molten drops of metal are flung off and solidify around the wear track, where they are clearly visible.

These speeds are reached in the sliding of the driving bands of shells as they are fired out of a gun barrel. They also occur in equipment known as SNORT used to test the aerodynamic behavior of various aircraft components. The components are mounted on a sledge which is fired by a rocket to speeds of up to 2,000 m.p.h. The sledge slides on guide rails on iron slippers which provide molten-film lubrication so that both the friction and wear are very small. Force-measuring devices are attached to the component and the forces experienced during travel are telemetered back to a base station. Evidently

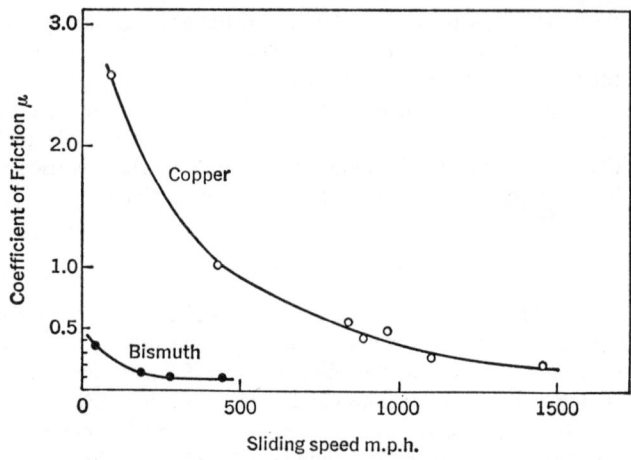

Figure 42 *Friction of a steel ball rubbing against copper and bismuth at extremely high sliding speeds. Modest vacuum. With copper the friction falls to a relatively low value at sliding speeds of 1,500 m.p.h. due to the formation of a thin molten surface film of copper. With bismuth, where the melting point is much lower, the drop-off in friction occurs at a correspondingly lower speed.*

the friction of metals at very high speeds can be much less than at normal speeds and the wear may also be very much less. This suggests that the problem of sliding metals at very high speeds may not be as difficult as we might have supposed. If only one could move fast enough one might ski on copper mountains.

Friction at Very High Temperatures

Recent developments in aircraft engines and nuclear power stations have made it more and more desirable to produce materials which will slide or rub over one another at very high temperatures without the danger of seizure or mechanical failure. Steels and other more conventional alloys cease to be useful materials mechanically at temperatures above about 800°C. At these and higher temperatures it is necessary to use carbides, borides and nitrides of silicon, tungsten, boron, vanadium and similar materials. Such materials are known as refractory solids, i.e., "stubborn" materials, which can withstand high temperatures. In the technical literature they are also called "hard metals." This is misleading since they are not metals; they are not even alloys, they are chemical compounds. They are usually very hard at room temperature and are often too brittle to be of practical use. However, they have very high melting points (of the order of 3000°C), so that they are able to maintain their strength up to 1500°C or even higher.

Some typical friction results on carbide specimens in vacuum or in an inert atmosphere are shown in Figure 43. Com-

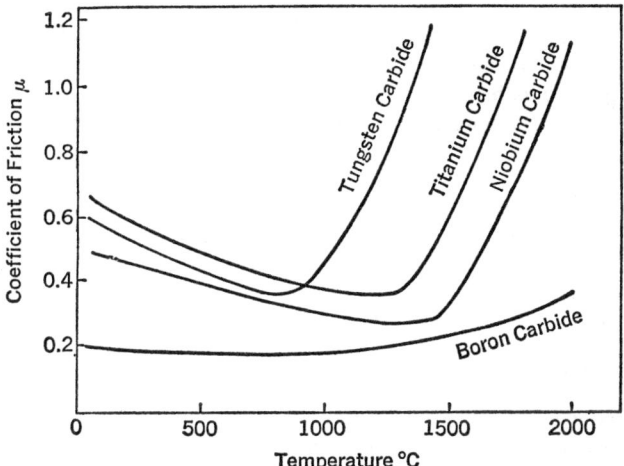

Figure 43 *Friction of a number of carbides sliding on themselves at very high temperatures. Around 1500°C there is a marked rise in friction due to a large increase in the diffusion of surface atoms: this strengthens the friction functions. Boron carbide does not show this effect to the same degree as the other carbides.*

pared with metals under the same experimental conditions these materials have a much lower friction, especially below 1500°C. However, above this temperature there is a marked increase in friction. This is partly due to a softening of the materials and an increase in their ductility. An additional factor is the increase in diffusion across the interface which strengthens the adhesion between the surfaces.

It is natural to consider whether it is possible to lubricate such materials when sliding at elevated temperatures. Conventional lubricants are completely unsuitable, for they will decompose chemically at temperatures of only a few hundred degrees C. However, it should be possible to achieve some measure of lubrication by exposing the surfaces to a gas or vapor which can react to form an appropriate surface film. For example, as was pointed out in an earlier chapter, it is possible to lubricate molybdenum by exposing it to hydrogen sulphide to form MoS_2 on the surface. Similarly it is possible to lubricate titanium metal by exposing it to iodine vapor, while chromium metal can be effectively lubricated by exposure to chlorine. All the compounds formed in these experiments have a layerlike or lamellar structure. Some typical results for pure metals are summarized in Table IV.

TABLE IV

FRICTION OF CLEAN METALS AFTER EXPOSURE TO VARIOUS GASES TO FORM LAMELLAR SURFACE FILMS

Metal	Coefficient of friction μ_s Clean	After reaction	Temp. at which lubrication deteriorates	Nature of surface film
Molybdenum	2.0	H_2S : 0.2	800°C	MoS_2
Uranium	1.2	H_2S : 0.4	700°C	US_2
Boron	1.0	N_2 : 0.7	1100°C	BN
Titanium	1.2	I_2 : 0.3	400°C	TiI_2
Chromium	2	Cl_2 : 0.2	700°C	$CrCl_3$

Recent experiments show that we can achieve a similar type of lubrication with some of the refractory compounds mentioned above. For example, molybdenum disilicide ($MoSi_2$) when sliding on itself has a friction of the order of 0.6. When exposed to H_2S it forms a surface layer of MoS_2 and the friction reaches a value of $\mu = 0.03$ at 700°C. Above this temperature the sulphide decomposes and the friction rises. Tungsten carbide (WC) and tungsten diboride (W_2B) behave in a similar manner. When exposed to H_2S a low friction of about $\mu = 0.03$ is achieved which is maintained up to about 950°C. For both the molybdenum and tungsten compounds the surface film has been shown to possess a lamellar structure.

With titanium compounds exposed to H_2S a surface film is formed but examination shows it to be the mono-sulphide —not the disulphide. It does not have a lamellar structure and although a reduction in friction is achieved ($\mu \cong 0.1$) it is not as low as with the molybdenum and tungsten compound.

These results, although of a preliminary nature, are very promising. They suggest that by the choice of appropriate vapors it may be possible to lubricate certain refractory solids up to temperatures exceeding 1000°C. It should, however, be remembered that H_2S, which is a very convenient vapor to use in the laboratory, is highly toxic and its use in industrial applications could be extremely hazardous.

VIII LUBRICATION

> Some people put on the ground planed boards because of their smoothness and smear them with grease, because the roughness that is on them is made smooth and so they move the burdens with smaller power.
> Heron of Alexandria (c. A.D. 60.)

It has long been known that liquids and greases reduce the friction between sliding surfaces. The ancients believed that this was because the liquid or grease filled the surface hollows and so presented a smoother surface for sliding. This point of view, which has the support of the philosophers of antiquity, was also the view of Amontons, Coulomb and engineers until beyond the middle of the nineteenth century. Conservatism and authoritative opinions are to be found in science as in other walks of life. For example, Gustave Adolphe Hirn (1815–90), who pioneered original research on lubrication during the early stages of the industrial revolution, had a paper reporting his work rejected by the French Academy of Sciences in 1847 on the grounds that "the results were contrary to the well-known principles of mechanics"—meaning in fact that the results were not in agreement with those of Coulomb.

The main achievements in lubrication are less than ninety years old. In this chapter we shall discuss some of the more important fields.

Hydrodynamic Lubrication

The modern period in lubrication starts with the classical work of Osborne Reynolds (1842–1912) on the behavior of shafts or journals rotating in bearings (see Figure 44). His theoretical analysis was based on some elegant experimental studies by Beauchamp Tower (1845–1904), who was concerned with the behavior of the axles and bearings in locomotive wagons. Tower had found that when the system was running effectively quite a high hydrostatic pressure was set up in the lubricant film (see Figure 45). A paraphrase of Reynolds' explanation, published in 1886, follows:

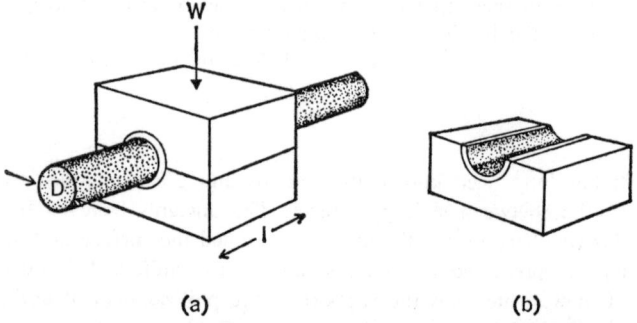

(a) (b)

Figure 44 (a) *Full bearing.* (b) *Half bearing. Nominal pressure in a bearing equals load divided by projected area of the bearing, i.e., $W/(D \times l)$. It is usually expressed in pounds per square inch (p.s.i.). With these bearings the working pressures usually range from about 100 to 1,000 p.s.i.*

If a shaft or journal rotates in a clockwise direction in a dry bearing it crawls up the right-hand side of the bearing as shown in Figure 46a. If, however, the bearing contains oil and the system runs properly the shaft is displaced to the other side of the bearing and drags a converging wedge of oil between it and the bearing surface. As a result there is an increase in velocity of flow as the lubricant passes through the narrowing constriction between the moving surfaces. Reynolds showed that because the liquid is viscous this produces

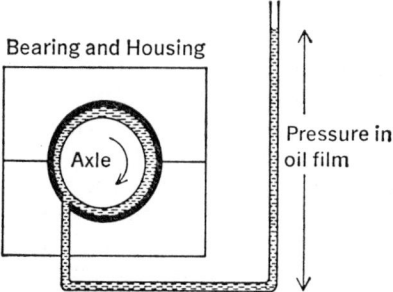

Figure 45 *Beauchamp Tower's experiment with a journal or shaft running in a well-lubricated bearing. He found that a high pressure was set up in the oil film, large enough to support the weight of the shaft.*

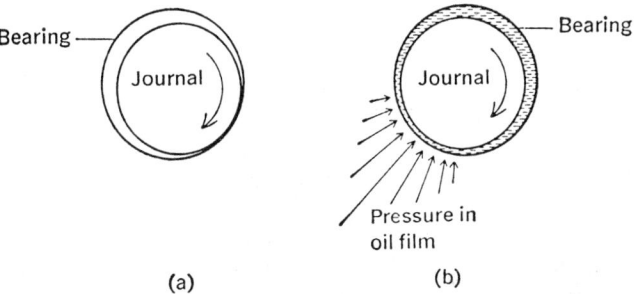

Figure 46 *Journal or shaft rotating clockwise in a bearing. (a) Unlubricated, the shaft climbs up to the right. (b) In the presence of a suitable lubricant the shaft moves to the left and drags a wedge of oil with it. The pressure developed in the oil film is due to two factors: first, the increase in flow rate of the oil as it enters the narrowing constriction and, secondly, the fact that the oil shows a viscous resistance to flow. Osborne Reynolds showed that under ideal hydrodynamic conditions the oil pressure could support the load and that all the friction arises from the viscous resistance of the oil.*

a build-up of liquid pressure in the oil wedge and this is, under favorable conditions, sufficient to keep the surfaces completely separated. The pressure distribution in the oil film is shown schematically in Figure 46b.

The word hydrodynamic conveys two ideas: a liquid (hydro) and relative motion (dynamic). There are in fact two other features which are essential in hydrodynamic lubrication. First, the liquid must be viscous and indeed the viscous properties are very important: second, the geometry of the surfaces must be such that as one surface moves over the other they must produce a convergent wedge of liquid. This occurs in a natural way when a shaft rotates in a bearing as discussed above. It can also be made to occur between two flat surfaces if one of them is slightly inclined to the other. This is used in what is known as the tilted-pad bearing and is illustrated in Figure 47. In what follows we shall for simplicity restrict our discussion to a shaft in a bearing.

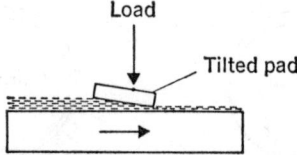

Figure 47   *Hydrodynamic lubrication in a tilted-pad bearing.*

In hydrodynamic lubrication the whole of the friction arises from shearing the lubricant film so that it is determined by the viscosity of the oil: the thinner (or less viscous) the oil the lower the friction. There is, however, another factor which places a limit to the lowest viscosity that is permissible, i.e., the distance of nearest approach. This distance gets smaller the higher the load on the bearing, the lower the speed and the lower the viscosity of the oil. If the viscosity of the oil is too low this distance may become smaller than the height of the surface irregularities: the film may then be penetrated with a consequent increase in friction and wear.

The great advantages of hydrodynamic lubrication are that the friction can be very low ($\mu \cong 0.001$) and, in the ideal case, there is no wear of the moving parts. Of course the

Lubrication

surfaces must be smooth, and the shaft and bearing must be well aligned and of the right dimensions permitting just the right amount of clearance. Further, allowance must be made for the fact that the frictional work goes into heating the oil and this reduces its viscosity. A thinner oil gives a reduced oil film thickness so that the danger of oil penetration increases with running time. There are two ways of coping with this. One is to have a large supply of oil which is continuously circulated through the bearing and cooled in a suitable reservoir or tank. The other is to use special additives, now commercially available, which make the viscosity of the oil far less sensitive to changes in temperature.

The main problems in hydrodynamic lubrication are associated with starting or stopping since the oil film thickness theoretically is zero when the speed is zero; there are also the complications which arise if the shaft has to cope with alternating loads as in the crankshaft of an internal combustion engine. In principle any hard strong material should serve effectively as a bearing for a system running under ideal hydrodynamic conditions. In practice local breakdown of the lubricant film may occur and this may lead to serious failure. For this reason bearings are made of special materials which can cope with the situation if shaft and bearing come into momentary, intimate contact.

Aerodynamic Lubrication

There is a recent extension of hydrodynamic lubrication which is interesting and of growing importance. It consists of using air or some other gas as the lubricant. The viscosity of air is 1,000 times smaller than that of a very thin mineral oil. Consequently the viscous resistance is very much less. However, the distance of nearest approach, i.e., the closest distance between the shaft and the bearing, is also correspondingly smaller, so that special precautions must be taken. To obtain full benefit from such aerodynamic lubrication, the surfaces must have a very fine finish, the alignment must be very good, the dimensions and clearances must be very accurate, the speeds must be high and the loading relatively low. If all these conditions are fulfilled extremely successful bearing systems can be made to run at very low coefficients of friction. They may also operate at very high temperatures since chemical degradation of the lubricant need not occur. Furthermore if air is used as the lubricant, it costs nothing.

Elasto-hydrodynamic Lubrication

In the arrangement of shaft and bearing considered in a previous section we assumed that the surfaces were perfectly rigid and retained their geometric shape during operation. We may now ask: what is the situation if the geometry or mechanical properties of the materials are such that appreciable elastic deformation of the surfaces occurs? Let us first discuss mechanical properties. Suppose a steel shaft rests on a rubber block. It deforms the block elastically and provides an approximation to a "half-bearing" (Figure 48a). If now a lubricant is applied to the system it will be dragged into the interface and, if the conditions are right, it will form a hydrodynamic film. However, the pressures developed in the oil film will now have to match up with the elastic stresses in the rubber. In fact the shape of the rubber will be changed as indicated in Figure 48b. This type of lubrication is known as elasto-hydrodynamic lubrication. It occurs between rubber seals and shafts. It also occurs, rather surprisingly, in the contact between a windshield wiper blade and a windshield in the presence of rain. The geometry of the deformable member, its elastic properties, the load, the speed and the viscosity of the liquid are all important factors in the operation of elasto-hydrodynamic lubrication.

With conventional journals and bearings the average pressure over the bearing is of the order of 1,000 pounds per square inch (p.s.i.). With elasto-hydrodynamic bearings using a material such as rubber the pressures are perhaps 10 to 20 times smaller. Let us now consider the other end of the pressure spectrum. In gear teeth, contact pressures of the order of 100,000 p.s.i. may be reached. Because the metals are very hard this may still be within the range of elastic deformation. With careful alignment of the engaging gear teeth and appropriate surface finish, gears can in fact run successfully

Lubrication

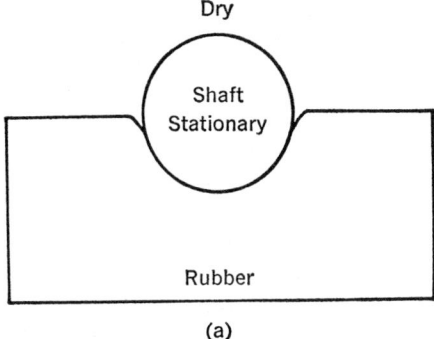

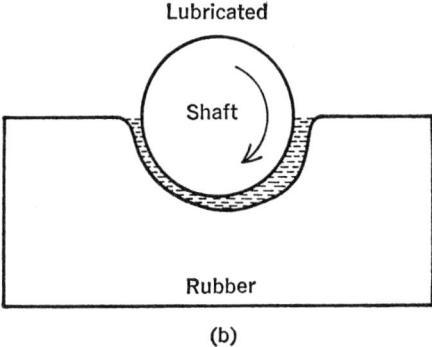

Figure 48 (*a*) *A shaft loaded on to a pad of rubber deforms it to form a "bed" which resembles a rigid half-bearing.* (*b*) *If the shaft is set rotating in the presence of a suitable lubricant a hydrodynamic film may be set up which can support the applied load. The pressures, however, will now be sufficient to modify, by elastic deformation, the shape of the rubber half-bearing. This is the simplest example of elasto-hydrodynamic lubrication.*

under these conditions using an ordinary mineral oil as the lubricant. If we calculate the thickness of the elasto-hydrodynamic film formed at such pressures we find that it is less than an atomic diameter. Since even the smoothest metal surfaces are far rougher than this (a millionth of an inch is about 100 atomic diameters) it seems hard to understand why lubrication is effective in these circumstances.

The explanation was first provided by the Russian engineer Dr. A. N. Grubin in 1949 and a little later (1958) in Britain by Professor Alfred William Crook. With most mineral oils the application of a high pressure can lead to a prodigious increase in viscosity. For example, at a pressure of 100,000 p.s.i. the viscosity may be increased 10,000-fold. The oil entering the gap between the gear teeth is trapped between the surfaces and at the high pressures existing in the contact region behaves virtually like a solid separating layer. This process explains why many mechanisms in practice operate under much severer conditions than the classical theory would allow. We may repeat our comment in Chapter III that in this field, at least, nature is much kinder to the engineer than he had supposed.

This type of elasto-hydrodynamic lubrication becomes apparent only when the film thickness is less than about 10 to 40 millionths of an inch. To be exploited successfully it implies that the surfaces must be very smooth (roughnesses of only a few microinches) and very carefully aligned. If these

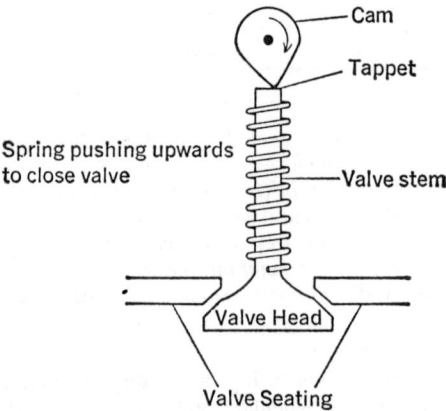

Figure 49 *Arrangement of a cam-and-tappet mechanism used to actuate the valves in an automobile engine. The contact pressures between cam and tappet are of the order of 100,000 p.s.i.*

precautions are taken systems such as gears or cams and tappets (Figure 49) can operate effectively at very high contact pressures without any metallic contact occurring. The coefficient of friction depends, of course, on the experimental conditions of load, geometry, speed, etc.; but generally it lies between about $\mu = 0.01$ at the lightest pressures and $\mu = 0.1$ at the highest pressures. The great success of elastohydrodynamic theory in explaining effective lubrication at very high contact pressures also raises a problem that has not yet been satisfactorily resolved: why do lubricants ever fail, since the harder they are squeezed the harder it is to extrude them? It is possible that high temperature flashes are responsible; alternatively the high rates of shear can actually fracture the lubricant film since when it is trapped between the surfaces it is, instantaneously, more like a wax than an oil. More work on this is needed.

It is clear that in this type of lubrication the effect of pressure on viscosity is a factor of major importance. It turns out that mineral oils have reasonably good pressure-viscosity characteristics. These materials have long been known and some of their properties were described by the ancients. In the sixteenth century, Georgius Agricola,* who wrote one of the first textbooks on mining and metallurgy, observed that since these liquids were extracted from rocks they should be called "petroleum" oils (*petra* = rock). The petroleum companies are exploiting a property of their lubricants that they did not know they possessed. Synthetic oils are not so fortunate and probably one of the reasons that they are less successful as lubricants is that their pressure-viscosity characteristics are not so satisfactory.

Boundary Lubrication

If the pressures are too high, the running speeds too low or

* Georgius Agricola (the latinized form of Georg Bauer) was born in Saxony in 1494 and studied philosophy, medicine and natural sciences in Italy. For the last twenty-odd years of his life until his death in 1555 he lived in Chemnitz, where he was both court historian and city physician. The first edition of his great work *De re metallica* was published posthumously in Basel in 1556.

the surface roughness too great, penetration of the lubricant film will occur. Contact will take place between the asperities: the friction will rise and—this is more important—wear will take place (Figure 50). As the running conditions are made more severe the amount of lubricant breakdown increases until the system finally scores or seizes so badly that it can no longer operate successfully. If we study the friction and wear of a system lubricated with a simple mineral oil we find that the viability of the system can be greatly extended if we add a small quantity of certain active organic compounds to the mineral oil. (We shall have a little more to say about these compounds in subsequent paragraphs.) These additives are present in such small quantities (less than 1 per cent) that they have a negligible effect on the viscosity of the lubricant. They function because they form surface films which are

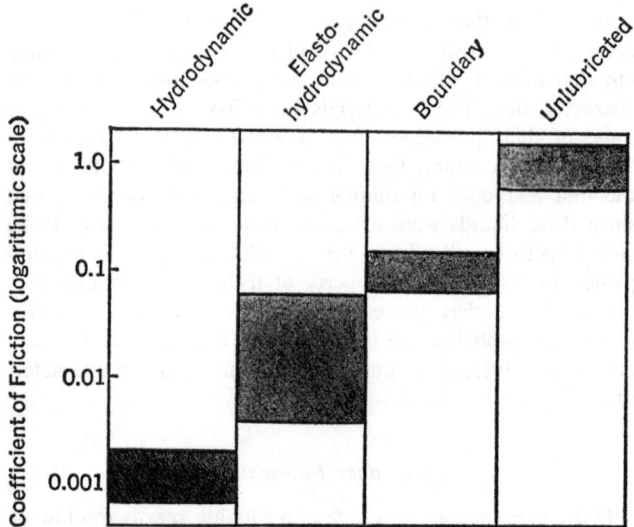

Figure 50 *Schematic drawing showing how the type of lubrication shifts from hydrodynamic to elasto-hydrodynamic to boundary lubrication as the severity of running conditions is increased. In boundary lubrication although the friction is much higher than in the hydrodynamic regime it is much less than for unlubricated surfaces.*

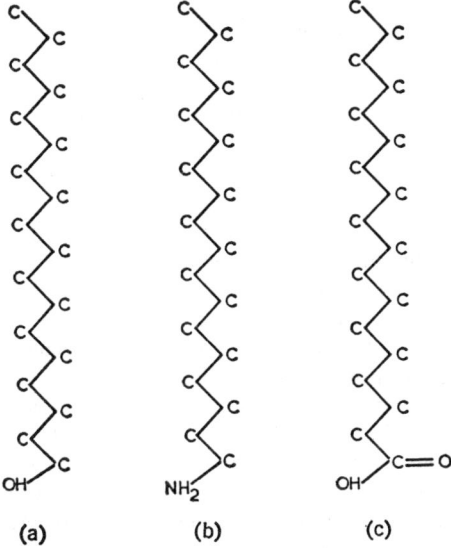

Lateral attraction between hydrocarbon chains

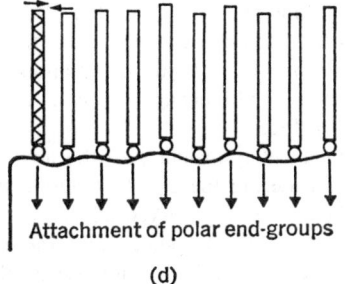

Attachment of polar end-groups

(d)

Figure 51 *Boundary lubricant molecules consist of a long backbone of carbon atoms with an active polar end group. (a) With alcohols the polar end group is OH; (b) with amines the polar end group is NH_2; (c) with fatty acids the polar end group is COOH. (d) Such molecules readily adsorb on a metal surface to form an oriented monolayer. The polar end groups are attached to the surface and there is strong lateral attraction between the chains.*

strongly attached to the metal surfaces. Although they are generally only one or two molecules thick they are able to prevent metal–metal contact. The type of lubrication provided by these last boundary layers is known as boundary lubrication.

Boundary lubrication is best studied by sliding surfaces over one another at extremely low speeds and very high contact pressures so that the incidence of hydrodynamic or elastohydrodynamic lubrication is reduced to a minimum. It is then found that the best boundary lubricants are long chain molecules with an active end group, typically an alcohol, an amine or a fatty acid. Representative molecules of these types are shown in Figure 51. They consist of a hydrocarbon backbone of carbon atoms and an active end group. With the alcohol the end group is OH; with the amine it is NH_2; with the fatty acid it is COOH, known as the carboxyl group. When such a material, dissolved in a mineral oil, meets a metal or other solid surface the active end group attaches itself to the solid and gradually builds up a surface layer as shown in 51d. This film is strongly attached to the surface. In addition, because the attraction between the chains themselves is strong it is very difficult to penetrate the film. If two surfaces covered with such films come into contact they tend to slide over their outermost faces. Some penetration may occur but it is far less than would occur if only a mineral oil were present on the surface.

We may gain a better insight into the action of boundary lubricants by considering the effect of temperature. Consider the frictional behavior of a curved slider passing over a flat surface. We make both the slider and the flat of a non-reactive metal such as platinum and use as the lubricant a film of stearic acid a few molecules thick. We may easily deposit such a layer by dissolving a small quantity of the acid in a volatile solvent, that is, in a solvent (such as benzene) which easily evaporates. If a few drops of the solution are spread over the flat surface the solvent will evaporate and leave the acid in the form of a thin film. We carry out the friction experiment at a very slow sliding speed for two reasons: first, so as to avoid hydrodynamic effects and, second, to keep frictional heating to a very low level. At room temperature we find that the friction is fairly low, $\mu = 0.1$. As we heat the whole sliding system we find that μ remains roughly constant until the temperature reaches 69°C, which is the melting point of stearic acid (see Figure 52). The friction then rises; there is a marked increase in metal–metal contact and appreciable surface damage. If the experiment is repeated with reactive surfaces, say, copper, the initial friction is lower ($\mu = 0.05$) and the breakdown does not occur until about 120°C. This is the softening point (or melting point) of copper stearate. Clearly chemical reaction has occurred. The chemical compound formed *in situ* is well attached to the surface: it is a better lubricant than the original fatty acid, and because it has a higher melting point it remains effective up to a higher temperature.

The protective properties of the boundary film can be studied conveniently using radioactive tracers in the way described in Chapter IV. The slider is made radioactive. During sliding some penetration of the lubricant film takes place and at these regions adhesion occurs between the slider and the

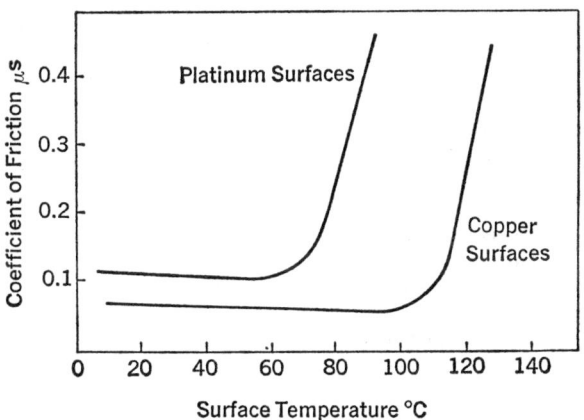

Figure 52 *Lubrication of platinum surfaces and of copper surfaces with pure stearic acid. On the non-reactive surface of platinum lubrication is effective up to the melting point of the acid (69°C). On the reactive copper surface lubrication is effective up to a higher temperature corresponding to the melting or softening of copper stearate (about 120°C).*

lower (non-radioactive) surface. A small fragment of the slider is then torn out and left attached to the lower surface. Since it is radioactive its presence and its amount may easily be detected using a sensitive Geiger counter. An alternative and more informative method is that of autoradiography. A photographic plate is laid over the flat surface and then developed. The blackening produced gives a picture of the distribution of the transferred fragments: in addition the intensity and diameter of the blackened spots provide a measure of their size. Some typical results of the transfer or pickup observed in the sliding of copper surfaces are given in Table V. It is seen that the copper stearate is by far the most effective lubricant.

TABLE V

FRICTION AND PICKUP ON COPPER SURFACE
Load 2,000 gm.
Room Temperature. Sliding speed 1 mm. per sec.

Condition of Surfaces	Coefficient of friction μ_s	Pickup 10^{-9} gm. per cm. of track
Unlubricated	1.2	20,000
Lubricated with:		
Silicone oil	1	10,000
Liquid paraffin	0.2–0.3	500
Solid paraffin	0.1–0.2	100
Solid alcohol	0.15	50
Fatty acid inadequate reaction	0.05–0.1	20
Fatty acid heavy reaction or Copper stearate	0.05	1

The Breakdown of Boundary Lubricating Films

We have already pointed out that the boundary lubricating film is most effective when it is in the solid form. A marked increase in friction and surface damage occur when the temperature of the surface exceeds the softening or melting point

Lubrication

of the film. The friction rises by a factor of about 5, the transfer or pickup by a factor of several hundred.† This stage is mainly associated with a weakening of the attraction between the chains themselves. The molecules still remain attached to the surface and are therefore still able to provide some protection to the surface.

At a somewhat higher temperature there is yet a further deterioration in lubricating properties. The friction rises by yet a further factor of 3 or 4 and the pickup or transfer by a further factor of several hundred. At this stage the friction and pickup are almost the same as for unlubricated surfaces (see Table VI). The molecules are now desorbed: by this we mean that although they are still present on the surface they have lost their attachment. Consequently wherever the surfaces come together the lubricant molecules are pushed away and intimate metal–metal contact is able to occur. With the best long-chain boundary lubricants desorption occurs above about 200°C. If the boundary film is covered with an oil in which it is soluble it may leave the surface at an even lower temperature.

TABLE VI

BREAKDOWN OF BOUNDARY LUBRICATION

Copper surfaces lubricated with stearic acid (heavy reaction)
Load 2,000 gm. Sliding speed 1 mm. per sec.

Temperature °C	State of film	Coefficient of friction μ_s	Pickup 10^{-9} gm. per cm. of track
20–100	Solid	0.05	1
100–120	Molten	0.2 to 0.3	500 to 1000
Above 150	Desorbed	0.9 to 1	10,000 to 20,000
Unlubricated		0.9 to 1	20,000

† At this stage the behavior is similar to that observed with an ordinary mineral oil such as paraffin oil at room temperature as may be seen by comparing row 3 in Table V with row 2 in Table VI.

The Action of Boundary Lubricants

The main function of a boundary lubricant is to interpose between the sliding surfaces a film that is able to reduce the amount of metallic interaction and that is itself easily sheared. The pickup is never entirely eliminated but extremely low values may be obtained when the film is in the solid state. A liquid, such as an ordinary mineral oil, is a relatively poor boundary lubricant. The best protection is provided by a solid film consisting of long chain molecules possessing the following properties: (a) strong attraction between the chains to resist penetration by surface asperities, (b) low shear strength to give a low friction, (c) high melting point so that it provides solid-film protection up to a high temperature. The best materials are the long chain alcohols, amines and fatty acids since these can be dissolved in small quantities in an ordinary lubricating oil and will then attach themselves to the metal surface. Fatty acids have the additional merit that, if the metal is reactive, they will react with the metal to form the metal soap. These materials are called "soaps" because ordinary toilet soap is the compound formed between a metal (sodium or potassium) and a fatty acid. They not only have desirable shear properties, they also have melting points considerably higher than that of the original fatty acid. For example, the melting point of stearic acid is 69°C; of copper stearate, about 120°C.

If we study the values quoted in Table V we see that a good boundary lubricant reduces the amount of pickup or transfer by a factor of 20,000 or more. This means that the amount of metallic contact is reduced by an enormous factor. In fact only the tip of an occasional asperity is able to penetrate the film and make metal–metal contact. If the friction of the lubricated surface was due to the shearing of the metallic junction so formed we should expect the friction to be reduced by a correspondingly enormous factor. In fact it is reduced by a factor of only about 24 (from $\mu = 1.2$ to $\mu = 0.05$). This implies that the major part of the friction comes from the shearing of the lubricant itself. That is why it is important that a good boundary lubricant should also be easily sheared.

Lubrication

These results have an important corollary. Suppose we compare two very good boundary lubricants, for example, the last two items in Table V. One may give the minute transfer or pickup shown as the last item in Table V. The other may give a transfer 20 times bigger: but this is still so small that the metallic contact involved will still only contribute a trivial part to the total friction. The friction will still be dominated by the shear strength of the lubricant film. In fact the coefficients of friction for the two lubricants may be indistinguishable from one another. Yet one protects the surfaces twenty times more effectively than the other. The conclusion is that with good boundary lubricants the coefficient of friction itself may not be a very reliable means of distinguishing between their protective properties. On the other hand, if the lubrication is poor a higher friction is generally associated with a higher transfer.

The transfer or pickup is, of course, a measure of the wear rate between the sliding surfaces. It is very small when the boundary film is solid. As the temperature is raised the attraction between the molecular chains is overcome by thermal motion (see Chapter III) and the film becomes molten. The asperities are more easily able to penetrate the film, with a consequent increase in friction and transfer. The film is still attached to the surface and so can still provide some protection. With increasing temperature this protection diminishes until finally the lubricant molecules are desorbed. They are no longer able to prevent metal–metal contact and the friction and pickup are similar to those observed for unlubricated surfaces. Clearly a good boundary lubricant should, in addition to its other desirable properties, maintain a strong attachment to the surfaces at high temperatures.

Greases

If a metal soap such as sodium or calcium stearate is mixed with an ordinary mineral oil the soap will act as a gelling agent and the material will turn into the soft deformable paste that we call a grease. Greases were known to the ancients, and Pliny the Elder in about A.D. 77 describes the manufacture of sodium soap and its use in grease. In most of these soap-base greases the soaps form long fibers, hundreds of molecules long, which interlock to form a sort of three-dimensional network: the oil is held within the cavities. Such greases are ideal lubricants under conditions where only a small amount of lubricating oil is required and also where it is desirable that the lubricant should remain put and not drip away. It is thus very appropriate for ball bearings, roller bearings, slideways, ball-and-socket mechanisms, etc.

The detailed lubricating action of greases is still the subject of dispute. It is well known for example that if undisturbed the grease will stay where it is placed; a small but definite force is required to move it. Once it is displaced it will flow like a very viscous liquid. If it is displaced or sheared vigorously it will flow much more easily until, ultimately, its flow properties resemble those of the base oil. These stages probably correspond to a progressive breakdown of the fiberlike soap network. If the grease is allowed to settle it will, to a large extent, recover its original properties. There is some evidence that some greases function by squeezing oil out of the network (the technical term is "bleeding") as and when required. If, of course, the situation becomes more severe the soap molecules themselves can take part in the lubricating process.

Not all greases contain soaps. A large class contains claylike particles of bentonite or silica. These have certain advantages because they can often be compounded to cover a wider temperature range without becoming unstable. However, the upper temperature at which a grease is effective is determined more by the properties of the liquid phase than by the gelling agent. For this reason it is not uncommon to include additives in the grease such as anti-oxidants, corrosion inhibitors and even extreme pressure or EP agents (see below).

Apart from their effectiveness as lubricants greases have two other attractive properties. They can often be used as a very simple means of reducing corrosion of metal parts. Secondly they can often act as quite effective seals in preventing access of dust and dirt to running mechanisms.

Extreme Pressure Lubrication

As we have just seen, the best boundary lubricants cease to be effective above 200–250°C. Furthermore, at these high temperatures oxidation of the lubricant may occur. If surfaces are to operate under more severe conditions other types of lubricants must be used. These are generally termed extreme pressure or EP lubricants but this is largely a misnomer. The main criterion is not the pressure but the temperature reached between the sliding surfaces.

EP lubricants usually consist of a small quantity of an EP additive dissolved in a lubricating oil, usually referred to as the base oil. The commonest additives used for this purpose contain phosphorus, chlorine and sulphur. In general these materials function by reacting with the surface to form a surface film which prevents metal-to-metal contact. If in addition the surface film formed has a low shear strength it will not only protect the surface, it will also give a low coefficient of friction. On the whole, chloride films give a lower coefficient of friction ($\mu = 0.2$) than sulphide films ($\mu = 0.5$) but the detailed behavior depends on the metals forming the sliding pair. On the other hand, sulphide films are generally more stable, are unaffected by moisture and retain their lubricating properties up to very high temperatures.

Although EP additives function by reacting with the surface they must not be too reactive, otherwise chemical corrosion may be more troublesome than frictional wear. They should only react when there is a danger of seizure, usually heralded by a sharp rise in temperature. For this reason it is often an advantage to incorporate in a lubricant a small quantity of a fatty acid which can provide effective lubrication at temperatures below those at which the additive becomes reactive. The behavior we are considering is shown schematically in Figure 53, where the coefficient of friction is plotted against the temperature. Curve I is for paraffin oil (the base oil) and shows that the friction is initially high and increases as the temperature is raised. Curve II is for a fatty acid dissolved in the base oil: it reacts with the surface to form a metallic soap which provides good lubrication from

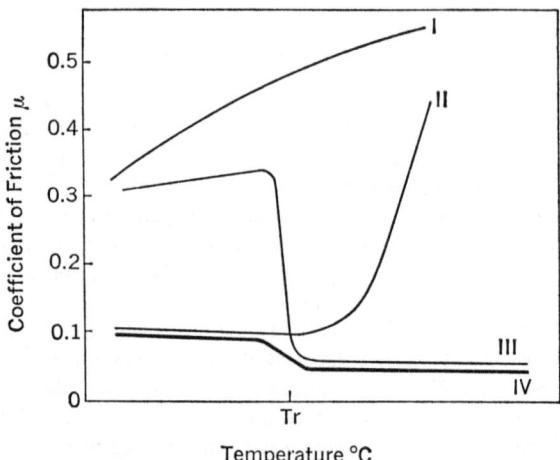

Figure 53 *Graph showing in a schematic way the frictional behavior of metal surfaces lubricated with: Curve I—paraffin oil; Curve II—fatty acid; Curve III—an EP lubricant which reacts with the surfaces at temperature T_r; Curve IV—a mixture of EP lubricant and fatty acid. The fatty acid provides effective lubrication at temperatures below that at which the EP additive reacts with the metal. At higher temperatures, when the fatty acid ceases to be effective the EP lubricant takes over.*

room temperature up to the temperature at which the soap begins to soften. Curve III is for a typical extreme pressure additive dissolved in the base oil: this reacts very slowly below the temperature T_r so that in this range the lubrication is poor while above T_r the protective film is formed and effective lubrication is provided up to a very high temperature. Curve IV is the result obtained when some fatty acid is added to the EP solution. Good lubrication is provided by the fatty acid below T_r, while above this temperature the greater part of the lubrication is due to the additive. At still higher temperatures, a deterioration in lubricating properties will also occur for both curves III and IV.

As we saw in Chapters IV and VII there are other ways of exploiting EP lubrication. For example, films of low-friction lamellar compounds may be formed *in situ* on many types of surfaces if the appropriate chemicals are used. Sometimes the chemical is a gas (H_2S, Cl_2, I_2), sometimes a liquid. A recent innovation in the lubrication of titanium is the use of a liquid containing active iodine: this reacts with the titanium to form a low-friction surface film of titanium iodide. Again recent work, especially in Russia, has shown that oxygen dissolved in mineral oils can play a vital part in preventing the seizure of heavily loaded friction surfaces by reacting to form surface oxide films. As most mineral oils inevitably contain some dissolved air or oxygen they possess a natural EP additive which must surely be the cheapest EP additive ever devised. It is a salutory thought that two of the most important attributes of ordinary mineral oils, their pressure-viscosity properties and the air that is dissolved in them, are unplanned gifts of the petroleum industry to the engineering profession.

IX ROLLING FRICTION; BALL BEARINGS; AUTOMOBILE TIRES; BRAKES; WOOD PULPING

> Why is it that spherical and cylindrical forms are easier to move? . . . firstly because they have a very slight contact with the ground and secondly because there is no friction, for the angle is well away from the ground.
> Why is it that it is easier to convey heavy weights on rollers than on carts although the latter have large wheels and the former a small circumference? Is it because a weight placed upon rollers encounters no friction, whereas when placed upon a cart it has the axle at which it encounters friction?
>
> Aristotle (384–322 B.C.),
> *Mechanica,* Sections 8, 11

In the preceding chapters we have dealt with the friction that arises when one body *slides* over another. There is, however, a different way in which we can move one surface relative to another and that is by rolling. It is much easier to roll surfaces than to slide them, an observation discussed with characteristic clarity by Aristotle over 2,000 years ago. With hard materials the coefficient of rolling friction may be as little as 0.001.

The First Rolling Bearings

The use of rolling, as distinct from sliding, as a means of obtaining a low coefficient of friction, finds its greatest application in ball and roller bearings. The first known examples of this in a practical mechanism are the turntables dating back to Roman times discovered when Lake Nemi was drained by Italian archaeologists between 1927 and 1932. Enormous ceremonial barges were kept by the Roman emperors on Lake Nemi and these were also exposed when the lake was drained. It is not known if the turntables were used for capstans on the ships or as rotating bases of statues. The bearings consist of a turntable revolving on small wheels or balls arranged in a circle. The wheels or balls have short projections on each side which act as shafts or journals (known as trunnions): these are supported in small journal bearings so that the arrangement is not that of a true ball or roller bearing. In a true ball or roller bearing the balls or rollers are free to roll in a circular groove called a race without any shaft or trunnion to support them.

Probably the first complete ball bearing in the modern sense is that due to Charles Varlo. In 1772 he published a fascinating booklet with the engaging title *Reflections upon friction with a plan of the new machine for taking it off in wheel carriages, windlasses and ships etc. together with metal proper for the machine and full directions for making it. To which is annexed "Stonhenge" one of the wonders of the world unriddled, printed for the author, London.*

A drawing of his bearing, reproduced from his book, is shown in Figure 54. It consists of three main components: the inner race B, the outer race F and the balls between them. The term "race" is used to describe the surfaces around which the balls roll. In Varlo's bearing the races are cylindrical in shape. The inner race B has a projecting ledge C attached to each side to stop the balls from falling out sideways. Similar ledges E are attached to the outer race F for the same purpose. (This is the significance of the lower sketch of Figure 54 containing the words "Ledge to keep in the globe.") In modern ball bearings the races are not cylindrical but have grooves to hold the balls in place. In addition there is usually a cage of some other material to keep the balls apart and prevent them rubbing on one another (see below). In Varlo's pioneer experiment these bearings were attached to the rear wheels of his post chaise. The outer race was fixed to the

Rolling Friction

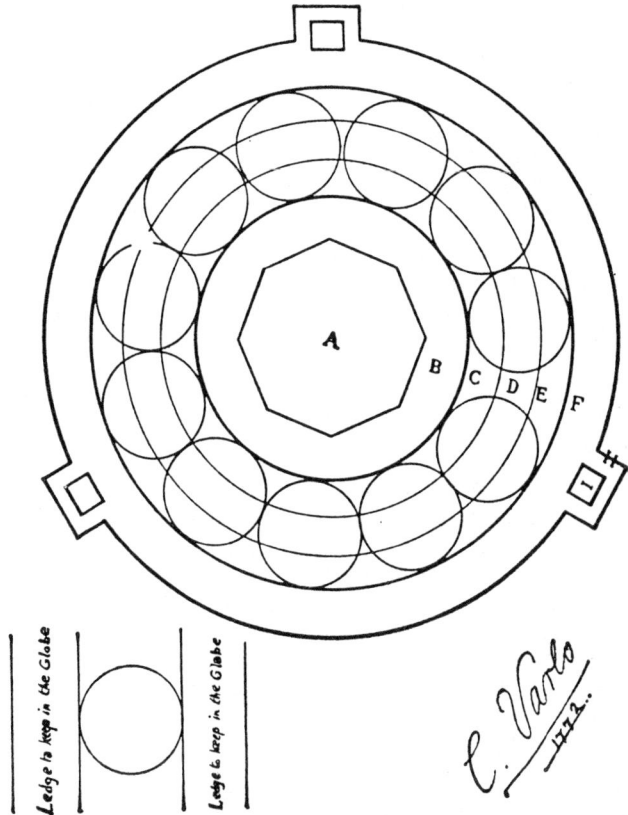

Figure 54 Ball bearing, designed and constructed by C. Varlo, Esq., 1772, and attached to his post chaise.

carriage itself and the inner race (in his terminology—the bush) to the wheel. He writes:

> In order to prove the machines, and to find out wherein the difference of the temper of the metal lay, etc., induced me to take long journeys in my own post chaise, which is a remarkably large heavy one. In it were two ladies, all

their luggage, and myself, drawn only by two horses. My first journey was from York to Liverpool, taking roundabout roads, and back again, through a great part of Derbyshire, to York.

In this journey I only had the inside bushes of the hind wheels fixed, where they have been ever since, viz. from the second of July 1772, to the twenty-ninth of October following; and am now on a long journey at Edinburgh; so that they have run at least 700 miles, which is no bad trial. They are not yet wore much above the sixteenth of an inch deep. . . ."

Of the temper of the ladies and how they wore, he says nothing. His experiments appear to have been disregarded and it is only in recent times that ball bearings which are essentially the same in design as Varlo's have come into use.

The Source of Rolling Friction

What is the source of rolling friction and why is it generally so small? Until recently most engineers considered that rolling friction arises from minute slip between the ball and the surface over which it rolls. However, recent studies both experimental and theoretical have shown that although such slip does occur it generally contributes only a small part to the rolling resistance. This is supported by the observation that lubricants, which greatly reduce sliding friction, have very little effect on rolling friction. We obtain a clue as to the source of rolling friction by considering what happens if a hard steel ball rolls over a softer metal such as lead or copper (Figure 55a). As it rolls along the ball displaces metal plastically around and ahead of it and produces a permanent groove in the metal surface. It is easy to demonstrate that the force required to displace the metal is almost exactly equal to the observed rolling friction. Thus the rolling friction is essentially a measure of the plastic work of deformation or grooving. We now see why lubricants have little effect on the rolling friction.

What happens then if we roll over an elastic solid such as rubber? No permanent groove is formed. Does this mean that the rolling resistance is zero? The answer is no. As the

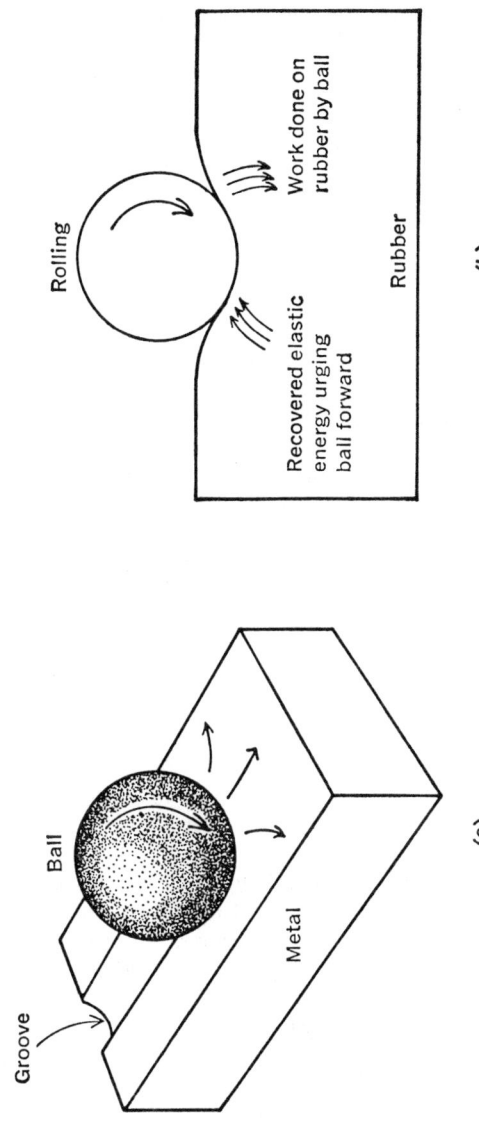

Figure 55 (a) When a hard steel sphere rolls over the surface of a softer metal it forms a permanent, plastically deformed groove. The rolling resistance arises from the work necessary to displace the metal from the path of the sphere. (b) On rubber no permanent groove is formed but energy is lost during rolling because of the imperfect elasticity (elastic hysteresis) of the rubber. This is the primary cause of the rolling friction.

ball rolls forward it deforms the rubber ahead of it and in doing so does work on it. The rubber recovers elastically and does work on the rear portion of the ball urging it forward (see Figure 55b). If we recovered as much energy in the rear portion as we expended on the front portion the net work required to roll the ball would be zero. But no material is ideally elastic. In the course of deforming and relaxing rubber, some energy is lost. This is known as elastic hysteresis (see Chapter III) and arises from the rubbing of rubber molecules over one another during the deformation process. Indeed elastic hysteresis is sometimes described as "internal" friction. For convenience we shall describe the fraction of the deformation energy lost by hysteresis as a percentage. With a very bouncy rubber the hysteresis losses are very small: only a few per cent of the deformation energy may be lost. On the other hand, with a very soggy rubber as much as 95 per cent of the deformation energy may be lost. This energy appears as heat within the rubber.

Ball bearings are made of hard steel and are so designed that the stresses are not high enough to produce plastic flow of the balls or races. Only elastic deformation occurs. For these steels the hysteresis losses are extremely small (less than ½ per cent), so that the rolling resistance is very small ($\mu \cong 0.001$). In practice the balls must be surrounded by a cage to separate them and prevent them rubbing on one another (see Figure 56). The cage friction is often far greater than the rolling friction. Lubricants are used to reduce the sliding friction between balls and cage and to prevent corrosion of the metal parts. They play little part in the rolling friction itself.

With rubberlike materials the rolling resistance can be very much larger. For example, with a ball loaded so heavily that it is half buried in the rubber and using a soggy rubber of high hysteresis loss a rolling resistance equivalent to a coefficient of $\mu = 0.3$ can be achieved.

Before leaving this we may ask a simple, somewhat embarrassing question. What is the role of adhesion at the regions of contact when rolling takes place? The first point to observe is that the rolling process imposes much gentler deformation on the surface than the sliding process. Consequently break-

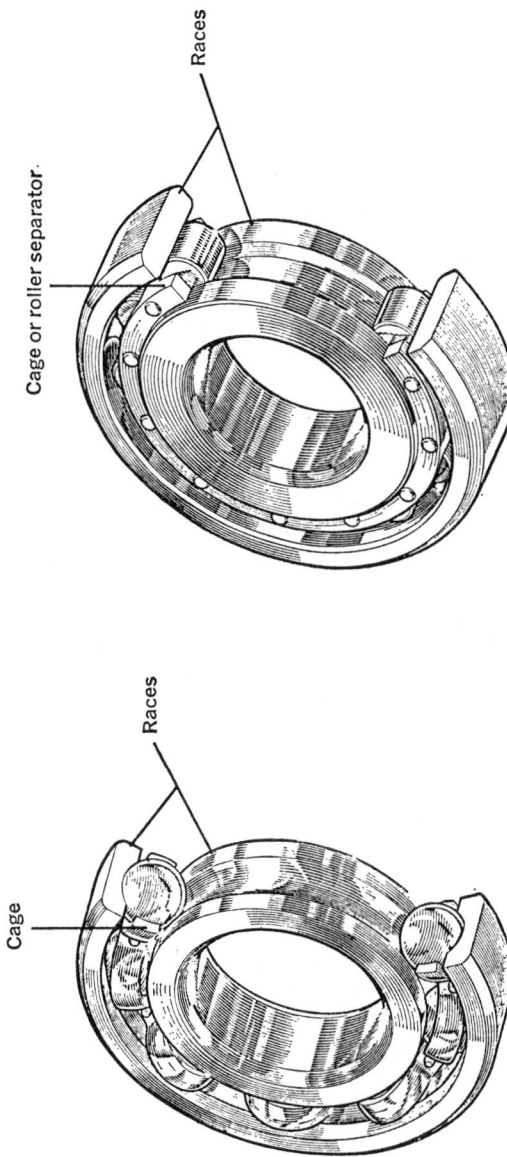

Figure 56 *Modern ball and roller bearings showing the outer and inner "race" and the cage to separate the balls or rollers.*

up of surface films is likely to be less marked so that strong adhesion is less likely. Secondly, even if adhesion does occur the junctions are peeled apart. This is a much easier process than shearing, as one may at once realize by considering the problem of removing the peel of an orange by shearing it off. The overcoming of interfacial adhesion consumes only a small part of the total energy expended during rolling. Thus even if a lubricant reduces the adhesion this has very little effect on the total rolling friction. This is the basic reason why lubricants play so small a part in rolling friction.

We now consider two practical examples where deformation losses play an important part in the frictional process.

The Friction of Automobile Tires

Automobile tires are relatively complex structures. They consist of a rubber carcass supported on polymeric or wire cords and a tire tread which rolls or slides over the road surface. In free rolling the tire is deformed as it meets the road surface and recovers as it leaves. If there is negligible slip between tire and road the energy loss is not large. For a standard car the equivalent coefficient of friction arising from the losses in the tire is of the order of $\mu = 0.01$ to 0.03. This consumes a small but finite amount of fuel. If the tire is made of a rubber with a higher hysteresis loss than the standard rubbers used, the rolling friction is larger and there is a larger power loss. This is, however, a small price to pay for comfort and safety.

The tire serves two main purposes: first, as a shock absorber in passing over rough road surfaces; second, as a means of gripping the road when accelerating, decelerating or cornering. We shall deal with the second purpose.

When a tire rolls freely over a highway the adhesion between the tire and road surface is of secondary importance. As we saw above the rolling friction is determined by other factors. If, however, we try to speed up, slow down or steer the automobile, slip occurs on a small scale between tire and road. In that case adhesion is of great importance. Its importance becomes more apparent if we apply the brakes violently and lock the wheels for then pure sliding occurs. Clearly if

Rolling Friction

the sliding friction is high we can brake the automobile in a short distance. Although it is not quite so obvious, the sliding friction is equally important in accelerating, decelerating and cornering. If the sliding friction is high these processes are relatively safe and the automobile is under good control. In fact, under normal conditions the sliding friction of tire on road is very high ($\mu = 1$ to 2). The position is very different if the road is wet* or covered with an oil slick. The sliding friction may be so small that we may find it difficult to brake or corner safely. We may indeed find that the automobile skids out of control. How can we improve the skid resistance of an automobile on such a road surface. The following laboratory experiment suggests one way in which this may be achieved.

Suppose we slide a hard steel ball over the flat surface of a clean strip of rubber. The adhesion and friction are very high (Figure 57a). Suppose now we lubricate the rubber with an effective layer of soap or grease so that all adhesion between the rubber and the steel is eliminated: the friction falls drastically. Part of the friction is due to the force necessary to shear the thin lubricant film and this may be very small indeed. The major part arises from deforming the rubber as the ball slides over the surface (57b). Indeed, the force needed to slide the ball over the rubber (when the surfaces are really well lubricated) is almost equal to the force required to *roll* the ball over the surface since it arises in both cases from elastic hysteresis losses in the rubber. This is shown in Figure 58. We are left with the striking conclusion that, in sliding, the friction may be largely determined by the bulk properties of the solids. It is very easy to demonstrate this. Instead of sliding a hard ball over a piece of rubber we can slide a flat piece of rubber over a rough surface; a convenient surface is the pimply type of glass sometimes used in bathroom windows (there is one very useful grade which has the romantic trade name of borealis glass). We now smear the glass with a pasty soap solu-

---

* If the road is covered with a thick layer of water (say, after a heavy cloudburst) the tires may ride up on the water in a manner resembling hydrodynamic lubrication. This is known as aquaplaning. The sliding friction between tire and road is then exceptionally low and control of the automobile extremely poor.

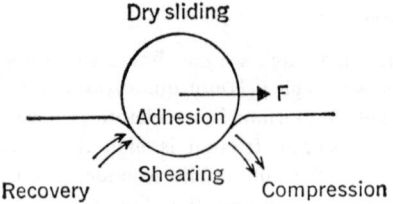

$F = F \text{ adh} + F \text{ deformation}$

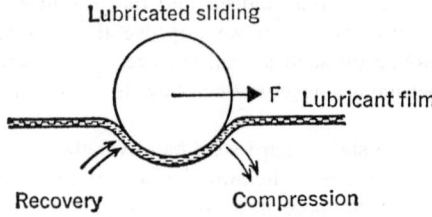

$F = F \text{ deformation}$

Figure 57 (a) *When a hard steel ball slides over a clean rubber surface the friction is dominated by the adhesion between the surface.* (b) *When the surfaces are thoroughly lubricated the friction is dominated by the deformation of the rubber as in rolling (compare Figure 55b).*

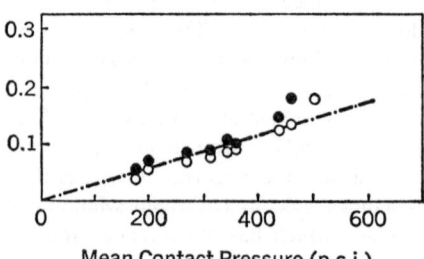

Figure 58 *Coefficient of friction against contact pressure for:* ○ *hard sphere rolling on rubber;* ● *hard sphere sliding on rubber lubricated with soap;* ·—·—·— *theoretical relation assuming the friction in both cases is due to elastic hysteresis losses. (Sliding speed 1 cm./sec.)*

Rolling Friction

tion, and slide our rubber specimen over it. The projections on the glass produce transient grooves in the rubber and energy is lost by elastic hysteresis. What do we in fact find? We find that soggy rubbers with high hysteresis losses have a higher friction than bouncy rubbers exactly as we predicted.

This may be carried a stage further. Some time ago Mr. Cyril George Giles and Miss Barbara E. Sabey of the Road Research Laboratories in Britain carried out some experiments on the friction of hard projections sliding over a flat strip of typical tire rubber. They found that above a speed of

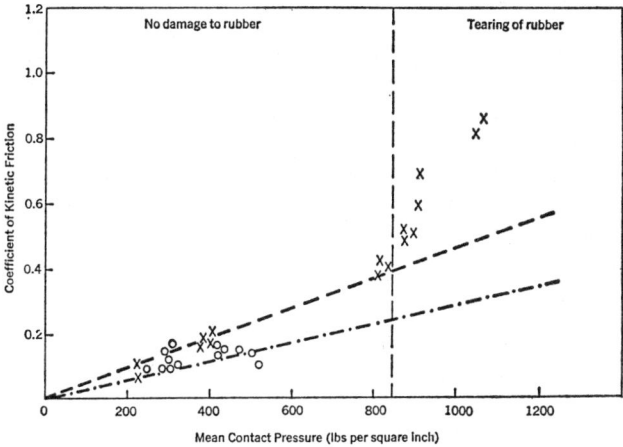

Figure 59 *Results obtained by Miss Barbara E. Sabey for hard spheres* ○ *and hard cones* X *sliding over rubber lubricated with water at a sliding speed of 6 m.p.h. The broken lines are theoretical results calculated on the assumption that the friction is due to elastic hysteresis losses. The agreement is very good except for the more pointed cones. These give a higher contact pressure, penetrate the water film and actually tear the rubber surface.*

a few miles an hour, water itself could provide almost perfect lubrication. Under these conditions of poor adhesion the friction was again largely determined by hysteresis losses in the rubber (see Figure 59).

There have hitherto been two approaches to the problem

of increasing the friction of tires on wet roadways. The first is that of the tire manufacturer who breaks up the surface of the tire into small areas with a typical tread pattern so that the water can be more easily squeezed out from between the tire and the road. In addition fine cuts are often inserted so that the tire itself acts as a sort of windshield wiper and wipes away the water, so ensuring good contact between road and tire. This approach can produce quite high friction on wet roadways. A second approach is that of the highway engineer who has found that certain types of road surfaces can penetrate the water film and so give a much better grip between road and tire. This type of approach has been pioneered in Great Britain by the Road Research Laboratories; they have shown that certain types of grits incorporated in the road surface can greatly increase skid resistance.

A third approach is suggested by the newer work just described. For if the tire slides or skids on a wet or oily roadway, that is, if the adhesion is very small, quite a large part of the friction will arise from the deformation of the tread

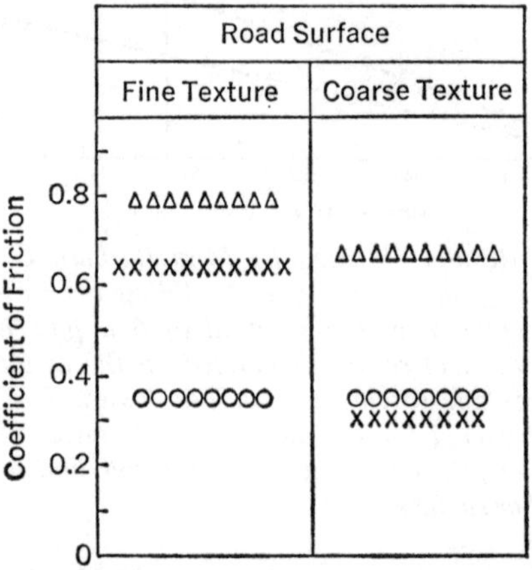

Figure 60 *Coefficient of friction of tires on wet roads for fine-textured and coarse-textured road surfaces.* ○○○ *smooth tread;* XXX *patterned tread;* △△△ *patterned tread of high hysteresis-loss rubber. The last gives a greatly increased friction, especially on the coarse-textured road surface.*

itself by the projections on the road surface. It follows that a tread with high hysteresis losses will have a higher friction, that is to say, a better skid resistance than a tread of more conventional bouncy rubber. Experiments show that such an improvement does occur (Figure 60). In practice the problem is far more complicated than we have indicated here but it is an interesting thought that one may be able to "build in" a higher friction into the tire itself. Indeed, high hysteresis-loss tires are now available commercially and are generally regarded as providing increased skid resistance.

A Few Words on Brakes

The previous sections have referred so much to skidding as a result of braking that we may interpose here a few comments on the action of brakes. Braking is one of the few mechanisms where a high friction is required. The brake material is pressed against a metal disk or drum and the frictional force developed is used to slow up or stop the moving parts. In most cases the disk or drum is of steel or a "fine-grained" cast iron. The brake material, however, varies greatly with the type of application. In railway vehicles the brake material is usually cast iron, and is pressed against the steel rim of the locomotive or carriage wheels. The friction is of order $\mu = 0.2$. The wear life is fairly protracted (about one year) and the cost of the brake material is low. In recent years automobile-type brake materials have been introduced.

In automobiles the brake material is usually made of cotton or asbestos fibers bonded together by (or impregnated with) a suitable resin. These materials give a coefficient of friction of the order of $\mu = 0.4$. The asbestos-based materials will usually withstand more severe conditions than those based on cotton because asbestos is highly stable chemically. Consequently, during braking, it will better survive the very high temperatures that may be generated by the friction process itself.

The major defect of the resin-bonded brake material is that the friction may not remain constant. If the brake is applied for a prolonged period the frictional heating may soften or melt the resin (or some other low-melting-point constituent) and smear the molten film between the rubbing surfaces. The friction then falls. If the brake is released and allowed to cool, the friction will return to its original higher value. This tendency of the friction to decrease as braking proceeds is known as "fade." Since fade implies a loss of control in the braking process it is desirable to reduce it to a minimum. In the last few years improved resin-impregnated materials with far less fade have been developed and these are finding increased use in automobile brakes.

In aircraft the braking of a modern jet liner involves the generation of enormous amounts of frictional heating. The brake disks may actually glow red-hot (temperature about 700°C). Clearly it is essential to secure a high and constant friction. For this purpose it is usual to use brake materials made of sintered metal. They are relatively good conductors of heat, give a coefficient of friction of about $\mu = 0.3$, have little fade and, in spite of the severity of the operating conditions, can function effectively for several dozen applications. They are, however, about three times as expensive as conventional brake materials and for this reason are not likely to find wide application as automobile brakes.

Another important use for brakelike materials is in clutches, where the friction between the clutch plate and the clutch facing serves to transmit power from the engine to some other moving part of the machine. The clutch plate is usually of steel or cast iron; the clutch facing, of some suitable brake material. Indeed, for this purpose ordinary brake materials can be used. A recent development in clutch facings involves the use of sintered metals in *the presence of a lubricant*. The coefficient of friction is relatively low ($\mu = 0.05$ to 0.1) but the frictional *force* can be increased by increasing the force with which the clutch plate and the clutch facing are pressed together. This raises several practical difficulties. On the other hand, the sliding is never jerky, the wear is low and there is little fade.

The chief requirements of the ideal brake (or clutch) material are a high coefficient of friction, a minimum of fade, very low wear and very low price. Not all of these are mutually compatible.

Pulping of Wood

The second example where deformation losses play an important part is perhaps a surprising one. It concerns the pulping of wood. In this process a large stone wheel impregnated with very fine grits is pressed against the surface of the wooden specimen. The wheel travels very fast and there is a copious supply of water. If conditions are favorable good-quality wood pulp, that is to say, wood in the form of long thin fibers is produced. It used to be thought that the fibers were torn out of the wood specimen by the fine grits. Recent work by Dr. Douglas Atack of the Pulp and Paper Research Institute of Canada suggests a far more interesting process. It turns out that, under the operating conditions of successful wood pulping there is very little adhesion between the grits and the wood. Most of the frictional work is expended as hysteresis losses within the wood.

Now, a study of the deformation of wood by a hard projection shows that most of the energy dissipated as hysteresis does not appear at the interface itself but a little below it in the bulk of the wood itself. It is at this region that the maximum heating takes place and in the wood pulping process it occurs at a depth just about equal to the thickness of a wood fiber. The suggestion is that the grits produce intense heating below the surface, degrade the lignin and other gluelike materials which hold the fibers together and so loosen each fiber from its neighbors. There is a good deal of evidence to support this view and on the basis of this mechanism it has proved possible to improve on the performance of wood pulping machinery.

Apart from our digression on the action of brake materials the examples quoted in this chapter are all extensions of a single idea: that when a hard body moves over a softer one some energy is lost in deformation. In the rolling process this is the major source of the rolling resistance. In the sliding process this may also constitute an appreciable part of the sliding friction, particularly if the adhesion between the surfaces is small. Indeed, with elastic materials the hysteresis losses involved in the deformation process may dominate the sliding frictional behavior between well-lubricated or poorly adhering surfaces.

X WEAR

. . . a ring is worn thin next to the finger with continual rubbing. Dripping water hollows a stone, a curved ploughshare, iron though it is, dwindles imperceptibly in the furrow. We see the cobble stones of the highway worn by the feet of many wayfarers. The bronze statues by the city gates show their right hands worn thin by the touch of all travellers who have greeted them in passing. We see that all these are being diminished since they are worn away. But to perceive what particles drop off at any particular time is a power grudged to us by our ungenerous sense of sight.

Lucretius (95–55 B.C.),
De rerum natura, I

The Non-existent Laws of Wear

In most machines the friction is less important than the surface damage and wear. A recent survey shows that in British industry more research is carried out on wear than on any other branch of tribology. Nevertheless our understanding of wear is far less satisfactory than that of friction and our ability to predict how a particular mechanism will wear is extremely poor. Consider the friction in air between unlubricated surfaces. The coefficient of friction for the most diverse systems and materials does not vary by more than a factor of about 20, e.g., $\mu = 0.05$ for PTFE and $\mu = 1$ for clean metals. Unless our theory of friction is completely unrealistic we cannot be far out in assessing the friction between surfaces. On the other hand, the difference in wear rate between, say, polyethylene on steel, and steel on steel may differ by a factor of 100,000. It is much easier to make enormous errors in our predictions.

There are no simple laws of wear as there are for friction. On the whole, hard solids wear less than soft, though polyethylene wears less than steel. The wear generally increases with the distance rubbed, though this may not be true if appreciable back transfer occurs (see below). The wear generally increases with load, though this may not be true if more severe running conditions produce appropriate structural changes in the surface layer.

At this point we must make a distinction between surface damage and wear. Surfaces may adhere very strongly and during sliding there may be continuous transfer of material from one surface to the other. However, if this is a to-and-fro process the net loss of material from either surface may be very small. For example, if two clean copper washers are rotated over one another the surfaces will be heavily roughened by adhesion and shearing of junctions. Since the washers are identical we cannot distinguish between one and the other. If material is transferred from one washer a similar amount will be transferred back from the other. Thus the surfaces may be badly damaged yet, in the initial stages, the wear may be negligible. At a later stage, of course, the transferred fragments may become detached and become loose wear particles. The surfaces then begin to lose weight and we can say that wear is taking place.

Although we have made this distinction between transfer and wear it is generally observed that if the transfer is high the ultimate wear will be high. A safer generalization is that if the transfer is small the wear will be small.

Adhesive Wear

Adhesive wear arises from the shearing of the friction junctions. If the junction is weaker than the material on either side of it, shearing occurs in the interface itself. There will be little surface damage and little wear. This situation occurs if sliding occurs within the surface oxide layer. If the junction

Wear

is stronger than one of the metals, shearing will occur not in the interface itself but at a little distance within the softer metal. This may lead to an enormous increase in both wear and surface damage for only a small increase in friction. The harder surface gradually becomes covered with a thin transferred film of the softer metal so that ultimately the sliding is characteristic of similar metals. Even in this situation there may be occasions where the softer metal plucks a small portion of the harder metal out of the surface. Finally, if similar metals slide together the junctions are of the same material as the parent surfaces but the frictional process itself will harden them and increase their strength. Consequently shearing will rarely occur in the interface itself but on either side of it within the bulk of the metals. The surface damage will be very large although sustained forward-and-backward transfer may mean that the actual wear is small. The heavy damage produced when similar metals slide together is the main reason for avoiding such combinations in sliding systems.

We see from this simple picture that although all the junctions contribute to the friction not all of them contribute to the wear. This has led Dr. J. F. Archard to describe the wear process in terms of a factor k which represents the fraction of the friction junctions producing wear. If k is 1 it means that every junction involved in the friction process produces a wear fragment. If $k = 0.1$ it means that one tenth of the friction junctions produce wear fragments. For clean gold surfaces k is between 0.1 and 1. For clean copper surfaces k is between 0.1 and 0.01, i.e., of the friction junctions formed only one tenth to one hundredth produce wear particles. Evidently clean gold surfaces wear about ten times more rapidly than clean copper surfaces. The actual wear rate depends not only on the value of k, but also on the hardness of the solids (see below). However, in general the smaller the value of k the smaller the wear rate. Dr. Archard has studied the wear behavior of a wide range of materials and finds that for unlubricated surfaces the lowest value of k is obtained for polythene sliding on steel. In this case $k = 10^{-7}$, which means that of the junctions responsible for friction only one in ten million produces a wear fragment.

Some typical results are given in Table VII. As mentioned

above, the actual wear rate is determined not only by k but also by the hardness of the material. For example, although the k value of polythene on steel is ten times smaller than for tungsten carbide sliding on itself the wear rate observed under comparable conditions is ten times larger, i.e., tungsten carbide wears ten times less than polythene. This is because tungsten carbide is so much harder than the polymer; consequently, at a given load, the area of the friction junctions and therefore the volume of the wear fragments is considerably smaller.

TABLE VII

THE WEAR OF SOLIDS: WEAR OF A SLIDER
RUBBING ON A FLAT DISK
Some typical values of Archard's k value

Rubbing materials	Coefficient of friction μ_s	k
Gold on gold	2.5	0.1 to 1
Copper on copper	1.2	0.01 to 0.1
Mild steel on mild steel	0.6	10^{-2}
Brass on hard steel	0.3	10^{-3}
Teflon on hard steel	0.15	2×10^{-5}
Stainless steel on hard steel	0.5	2×10^{-5}
Tungsten carbide on tungsten carbide	0.35	10^{-6}
Polythene on hard steel	0.6	10^{-7}

This table shows that wear rates can vary by enormous factors although the coefficient of friction may vary by only a factor of 10 or so.

Mild Wear and Severe Wear

If the sliding conditions are gentle, unlubricated surfaces can often slide together for very long periods without seizure, but they undergo wear. With metals this often proceeds in the following way. First, there is the transfer of fragments from one or both surfaces; then the fragments which remain stuck to the surfaces oxidize in the atmosphere; finally the oxidized fragments are removed by further sliding, so exposing fresh surface on which the wear process can be repeated. The surfaces will continue to run under these conditions of "mild" wear for prolonged periods, and as a result of friction and wear they will acquire a surface roughness of about 20 millionths of an inch.

If the loading is too severe the surface oxides are penetrated, the individual friction junctions are much larger and the torn fragments are lumps of metal which are too big to oxidize right through. The severe wear which results may be 100 or 1,000 times larger than in the mild wear regime and the surface roughness may be 50 or 100 times worse. The high wear and the large surface roughness imply that these conditions are not very suitable for prolonged rubbing.

Effect of Environment

Oxygen can play a very important part in the friction and wear process. Usually oxygen reduces wear but if it should turn out that the removal of transferred fragments can only take place after the fragments have oxidized, the wear may actually decrease if oxygen is removed from the system. Again, if the metal oxides are very hard and the conditions favor abrasive wear (see below) the presence of oxygen may increase the wear of at least one of the surfaces.

With ferrous materials the nitrogen present in air may play an important part. For example, under severe rubbing conditions where surface heating may be very marked, metal nitrides which are very hard and protective may form at the rubbing interface. These may greatly reduce the wear.

Lubricants generally reduce the wear because they reduce the amount of metal–metal interaction. The individual wear processes may resemble those observed with unlubricated surfaces but they are on a greatly reduced scale. However, with some lubricants chemical reaction may lead to considerable loss of metal. Indeed with some EP additives, if they are too reactive, wear by corrosion may exceed all other forms of wear.

Effect of Speed

The main effect of speed is to increase the surface temperatures generated at the sliding interface. These high temperatures will increase the rate of oxidation and of other types of chemical reaction. They may also lead to metallurgical changes. For example, in the sliding of dissimilar metals, alloys may be formed at the interface as a result of diffusion of one metal into the other. Now, alloy formation generally occurs only between metals that have an "affinity" for one another. This implies that there will be strong adhesion at the interface accompanied by high friction and wear. For this reason it has been suggested that for low wear it is desirable to select metals for the sliding surfaces that do not form alloys: or if the metals do form alloys these should be relatively brittle so that large-scale plastic flow at the interface is impossible. Although there are many examples which support this view, there are others which do not. Many pairs of materials (for example, lead on steel, copper on diamond) do not form alloys yet show very strong interfacial adhesion. Alloy formation is a bulk property, adhesion is a surface property and they are not necessarily equivalent.

Apart from the possibility of producing alloy formation, high rubbing temperatures can produce reactions with the environment and lead to the formation of new metallurgical compounds. For example, steels can react with the oxygen in the air to form oxides, with the nitrogen in the air to form nitrides and with the carbon in the lubricant (if it is a mineral oil) to form carbides.

The other effect of surface heating is physical. If the surfaces are softened by frictional heating they become much more ductile and much easier to deform. In that case plastic deformation may be greatly augmented and there is a high probability that both the friction and wear will increase catastrophically. In some cases, particularly if the speed of sliding is extremely high, surface melting will occur without appreciable softening in depth. Under these conditions both the friction and wear will be very low.

Surface Fatigue

There is another form of wear that is dominant when interfacial adhesion plays an unimportant part in the frictional process. This occurs in well-lubricated sliding systems where there is repeated encounter of asperities. They may never make direct contact but they may impose high stresses on one another through the lubricant film. After a few million such encounters the asperities may get "tired," they will fatigue and pieces of material may fall out of the surface. A similar process occurs in rolling bearings; continuous rolling produces a local loading and unloading of the races and this may lead to fatigue of the surface and subsurface layers. A piece of the race will then fall out. This is known as spalling and is one of the main types of failure observed in rolling bearings.

Abrasive Wear

The other main type of wear is abrasive. This is the type best understood, since it involves bulk properties rather than surface properties. In essence it involves a hard particle which indents, grooves and then cuts material out of the surface. This occurs in grinding and in the action of emery paper. The grits must be harder than the metal being abraded; they must be firmly held in their backing and they must have pointed rather than rounded tips so that they cut rather than groove. If the grits fracture during abrasion it is an advantage if the newly formed grits are also pointed. Finally it is generally necessary to prevent the abraded material from sticking to the grits and clogging up the spaces between them. This is the main reason that lubricants are used. If frictional heating of the surface by the abrasion process is to be minimized it is convenient to combine the lubricant with a cooling fluid.

It is clear that in abrasion the harder the surface the less the penetration by the grits and the lower the abrasive wear. There may, however, be a limit to this. If the surface is made too hard it may also become too brittle. Cracking may occur around the contact regions and relatively large flakes of material may fall out of the surface. Under these conditions, wear and surface damage could be very severe.

In the sliding of metals the main causes of abrasion are hard particles that may be present in one of the surfaces (e.g., carbide or silica inclusions in steel), work-hardened wear fragments, or the formation of very hard oxide films. Some of these surface films can be extremely abrasive and may dominate the wear process. For example, tin oxide is harder than steel or iron oxide: this explains the surprising observation that under conditions favoring oxidation a tin slider produces heavier wear of a chromium steel surface than does a hard steel slider. Similarly aluminum oxide is far harder than magnesium hydroxide. This may be *one* of the reasons (not the only one) why pistons made of magnesium alloy produce far less scoring and tearing of cylinder liners than pistons made of aluminum alloy.

The Role of Dirt in Wear

So far we have used refined scientific and technical terms to describe the various factors involved in wear. In fact in a very large number of practical mechanisms the main source of wear is dirt. This may be simply a piece of metal swarf left in the oil pump, a piece of scale that breaks off a feed pipe or a wear fragment itself. More commonly it is dirt drawn in from the environment. This includes ordinary dust, which consists of hard silica particles, inhaled, for example, by the carburetor of an internal combustion engine and fed into the cylinder. For a reasonably clean road this amounts to about one pound every four hours if filters are not used. With jet engines the contaminant may be gravel from the air strip or birds sucked in by the fans. These rarely affect the bearings but can quite often strip the blades off the turbine. In the heavy iron and metal industries metal flakes and oxide scale can often completely ruin the lubricant system. In heavy grinding machinery the dust from the grinding process is often involved.

Keeping a bearing or some other sliding mechanism clean is often nine tenths of the problem of preventing excessive wear or premature failure. In lubricating systems a good filter

Wear 155

can make an enormous difference. It is surprising to learn that although extremely efficient filters are now available on the market their detailed action is not really understood.

For bearings and similar mechanisms a most valuable adjunct is the rubber seal which rotates with very little friction or wear and yet prevents the access of dirt. Modern seals can be extremely successful. Their main defect to date is that their life and reliability are not very predictable.

The Complex Nature of Wear

We see from the previous sections that wear involves several mechanisms. These include adhesion and shearing of friction junctions, oxidation of transferred fragments, fatiguing of surface layers, abrasion by oxidized films, by wear fragments or by dirt, and chemical corrosion. Each of these processes is fairly well understood in itself, yet wear as a general phenomenon is usually very complex. This is partly because the sliding conditions may change the nature of the surfaces, either by work-hardening them or changing their roughness, or by producing new phases or alloys at the sliding interface. In addition, slight changes in the running conditions may alter the importance of individual wear mechanisms or change their mode of interaction in an unpredictable way.

XI SUMMARY AND PROSPECT: THE TRIBOLOGICAL CHALLENGE

Ay, there's the rub.
William Shakespeare (1564–1616),
Hamlet

From the ticking of the clock in the morning, throughout our work and leisure until we clean our teeth at night, we are involved in tribology. Some of these tribological phenomena are of social and human importance, for example, the friction between tire and road or the action of a dentifrice. Many of them are of great economic significance and it is this aspect that we shall be mainly concerned with in the sections which follow.

The Past

In the past the conquest or exploitation of friction and wear was a matter of trial and error. Engineering has nearly always proceeded by using empirical results and extrapolating them to another range of conditions. Such an approach can sometimes be astonishingly successful even if it is based on false premises. For example, at the middle of the nineteenth century it was found that lead-base and tin-base bearing materials were far superior to wood or brass or steel. Isaac Babbitt (1799–1862) observed that these materials had a soft "matrix" with hard particles scattered through it. With the instinct of the true inventor he therefore set about developing better white metal alloys which would exploit these two characteristics. As a result he produced (in 1839 and onwards) a series of "babbitts," which represented the best bearing materials then available. But we now know that this structure is an accident of metallurgy. The crucial component is the soft matrix, the hard particles being quite unimportant. Many successful white metal compositions are now available which contain no hard particles at all. The intuition of Babbitt led to great improvements in bearing materials but his approach was misleading. For example, there is another large class of bearing materials—the copper-lead alloys—in which the matrix (copper) is the hard phase while the soft phase (lead) is smeared over it. Porous bronze bearings constitute another class of material in which the continuous phase is hard (the bronze) while the grease or polymer which fills the pores is extruded as a soft film over the bearing surface. Again, another type of bearing consists of continuous layers of one metal deposited on top of another. Evidently intuition is not enough, particularly if it is based on erroneous assumptions.

Nevertheless the process of trial and error, of success and failure has led to marked advances in numerous technological applications. The engineer in conjunction with the metallurgist discovered largely by empirical means the best sorts of steels for ball bearings, for gears, for axles; the best types of alloys to employ in cylinders and pistons; the right sorts of cast iron to use for piston rings; the most suitable materials for brake linings. The engineer and the chemist, again largely by trial and error, developed the optimum grades for lubricating oils, the best compounds for EP additives, the most effective formulations for greases. But this approach, however successful, is gradually reaching its limit.

The Present

The current period has been largely one of investigating and understanding the basic mechanisms of friction, wear and lubrication. We now know, more or less, why white metal bearings work: because they have a soft, deformable low-melting-point constituent. We know, more or less, how the

Summary and Prospect

copper-lead bearings function and the mechanism of polymer-filled bronzes. We recognize that hard steels are desirable wherever contact stresses are high in order to minimize plastic flow and fatigue. We understand not only the operation of classical hydrodynamic lubrication but also the action of elasto-hydrodynamic lubrication and the role of pressure-viscosity characteristics. We have a good insight into the action of boundary and EP lubricants.

Of course there are still many aspects that we do not understand. But, by and large, the understanding we have acquired during the last forty years is considerable and there is scarcely a single tribological problem in engineering that cannot receive some useful guidance on the basis of this understanding.

The Tribological Challenge Confronting Industry

Wherever we have surfaces in relative motion we have some form of tribological problem. In Great Britain a survey of these problems in industry was carried out for the Department of Education and Science by an independent committee headed by Mr. Peter Jost. It was published in 1966 by Her Majesty's Stationery Office and is known as the Jost Report. It shows that tribological mismanagement involves enormous losses to industry, much of which could be saved by wise management and an enlightened approach. The power losses which occur in machinery because friction is too high are generally not very important. The main exception is in the enormous electricity generating plants now in use, where the frictional loss in the bearings can be appreciable. For Great Britain as a whole Jost estimates that an improvement in bearing friction in these large generators could save something of the order of seventy million dollars per year. The major losses to industry do not arise from this source: they are due to the labor costs of maintaining machinery in running order, the cost of replacing malfunctioning parts and the additional losses due to shutdown when the machinery is temporarily out of service. Jost estimates a total involvement of over two billion dollars a year in Britain, so that a 20 per cent improvement could mean a saving of nearly half a billion dollars a year. Some estimates suggest even higher figures.

One of the most interesting (and encouraging) features of this situation is that most of the existing problems could be overcome or mitigated on the basis of present-day knowledge. A vast number are due to faulty design or bad choice of materials. The designer of a piece of machinery often leaves a space "somewhere or other" for a bearing and then finds that it cannot cope with the demands placed upon it. He then calls in the bearing manufacturer and the oil chemist and complains if they cannot remedy the situation. Sometimes the size of the journal and the location of the bearings are such that the machine sets up vibrations in the journal: the designer then wonders why his machine falls to pieces. Again, the gear manufacturer may not impart the right shape to the gear teeth; or he may leave them with a surface finish rougher than the thickness of the oil film that forms between them. Naturally the gears (and the manufacturer) will run into difficulties. An even more common and less excusable error is the wrong choice of materials. If lubricant breakdown occurs the sliding surfaces may wear very heavily or even seize. Once again the designer turns to the lubricant specialist to solve a problem that should never have arisen.

Part of the problem is managerial, part psychological. Many practical men have a fatalistic attitude towards wear, failures and breakdowns. They regard them as acts of God, or of the Devil, depending on their theological views. But whatever place we may accord to fatalism in our general philosophy of life it has no place in technology. It may make life more comfortable to the harassed engineer but it leads to technological stagnation.

The challenge is to persuade the designer and the manufacturer of machinery that, with increased awareness, intelligence and some contemplation, they could resolve a large part of their tribological problems on the basis of existing knowledge.

The Economic Challenge

Some solutions to tribological problems may be available but they may not be justifiable on economic grounds. Suppose

Summary and Prospect

some large domestic appliance (for example, a washing machine) costs $200 and the main bearing costs $20. Suppose the life of the bearing is two years. A new bearing may then be installed for another $20 plus the labor cost of the mechanic. The appliance can then run safely for another two years. But supposing an improved bearing could be incorporated initially when the appliance was built and that this could run safely for four years. Presumably the manufacturer could afford to do this if the new bearing cost less than $40. The cost of the appliance would now be $220 with a guaranteed life of four years. Provided appliance-users do not expect a new model every two years the appliance containing the new bearing should be commercially viable. But if the improved bearing costs $50 or the customer becomes tired of his appliance after two years the revised model is not a worthwhile economic proposition.

Suppose a standard automobile piston ring costs $1. Suppose an improved ring with twice the life costs $2. The cost, in this example, is just proportional to the life. But the work involved in stripping an automobile engine in order to fit new rings will cost far more than the few dollars expended on improved rings. Clearly, *if one can be sure* that the more expensive ring has a proportionately longer life it is well worth installing initially. Furthermore, the better ring *may* also produce less wear of the cylinder walls and so prolong the life of the engine as a whole. This must also be taken into account. But if the customer wishes to change his automobile every year, irrespective of the degree of wear of the pistons or the cylinders, the better piston ring is no longer justifiable economically. On the other hand, it may increase the resale value of the automobile.

Suppose a better piston ring is desirable but does not yet exist. Finance will be needed for research and even more for development. Who should finance this venture? The piston ring manufacturer, the cylinder manufacturer or the automobile manufacturer? And how much can the industry as a whole afford to spend on such a project? Ultimately, of course, the customer will have to pay for it.

Consider finally a large compressor used for circulating fluids in a large chemical plant. The compressor may cost

$500,000, although its main bearings may cost only $100. The compressor may fail for many reasons such as fatigue, corrosion or abrasion by dirt particles. Suppose in fact it fails because of bearing failure. The new bearing will cost $100 and the labor costs of fitting it, perhaps $200—total cost, say, $300. But if the compressor is out of operation for a day the chemical plant will lose production for a day. This may involve a loss of $20,000. If the compressor is out of commission for a week the company will lose over $100,000. Clearly it would be worth paying $1,000 or more for a bearing if one could guarantee improved performance. But who can carry such a guarantee? The bearing manufacturer? Only if one could prove that failure is due to faulty materials. But failure is often the result of a number of interacting causes. No such guarantee is practical. Again if the compressor breaks down, not only is production halted. Other complex machines may be subjected to unusually heavy stresses and they in turn may fail. Should, then, the industry have a second, "standby" compressor to take over if the main compressor fails? Here again questions of cost, of life, of obsolescence are all involved.

Clearly economic factors, in the successful solution of tribological problems, are very important and are of great complexity. Nevertheless the phrase "not practical economically" is an excuse behind which poor management often hides.

The Tribological Challenge Confronting the Scientist: Materials

We first turn to problems concerning the sliding materials. As we have attempted to show in this book, we have a fairly good idea of the mechanism of friction between solids in terms of adhesion and deformation at the regions of real contact. Nevertheless it is impossible to predict the actual value of the coefficient of friction between a given pair of unlubricated surfaces in terms of their more basic physical properties. The cleaner the surfaces the more difficult the prediction, for we are then confronted with problems of the ductility of the junctions and, in particular, large-scale junction growth. If the surfaces are not clean the prediction is easier since junction

Summary and Prospect

growth is small. Fortunately most engineers work with dirty surfaces but even under these conditions the friction depends in a very complex way on surface finish, thickness of contaminant film, load and speed. It would be a great advance if we could tabulate the friction between various materials in a form that could be used reliably by the design engineer to cover the various operating conditions occurring in practice.

An associated problem concerns the ease with which different materials stick to one another. For example, clean gold will stick strongly to almost any other clean surface. Diamond will not stick readily to tungsten carbide even if the surfaces are clean. Poor adhesion in the latter case is partly due to the role of elastic stresses beneath the contact zone. Nevertheless other, rather elusive factors are involved. This has a relevance not restricted to the behavior of clean metals in a high vacuum or in outer space. It is also probably relevant to the behavior of sliding surfaces in the presence of a lubricant. If the lubricant film is maintained the nature of the underlying surfaces is relatively unimportant. But if for any reason the lubricant fails the behavior of the system may depend crucially on how the underlying solids interact when they come into intimate contact. It has long been known that certain metallic combinations will slide successfully under adverse conditions: they may "stick" a little and then "heal," whereas other combinations will fail very easily. In the internal combustion engine this type of failure is known as scuffing: it is still not clear whether scuffing is determined primarily by surface finish, lubricant inadequacy, high surface temperatures or the basic "stickiness" of the underlying materials.

Again, the action of dry bearings is fairly well understood. These usually consist of a porous metal backing (which provides strength and good heat conduction) and a low-friction material incorporated into the pores. A major group of these bearings uses Teflon as the low-friction material. Yet in order to reduce the wear of the bearing as a whole to acceptable limits a small quantity of lead, lead oxide or cadmium oxide has to be added. We do not understand why this is so effective: if we did we might be able to find better ways of achieving improved performance.

Another area of challenge concerns wear. We understand

individual wear mechanisms, yet it is impossible to predict the wear of any specified piece of machinery because there are too many interacting wear processes. The engineer would like to be able to put numbers to the wear properties and the wear life of solids rubbing together under various conditions. It may be impossible ever to do so, so that the challenge is likely to remain with us. One thing is clear: certain combinations give disastrously large wear rates and should never be employed by an intelligent engineer.

Wear of course may be mitigated by using special coatings. Experience and intuition suggest that hard coatings are the best for this purpose. Yet recent work suggests that in some cases soft pliable coatings can be far more effective. There are some ideas as to why this may be so but the subject needs far more study and clarification.

Lubricants and Lubrication

We may now turn to problems concerning lubricants and lubrication. The classical study of hydrodynamic lubrication was initiated by Osborne Reynolds almost a century ago. It has well stood the test of time. We not only understand how the liquid film generates a pressure between the moving surfaces, we can actually *calculate* it and determine how viscous the oil should be, how close the clearance between shaft and bearing should be if the oil film pressure is to be able to keep the surfaces apart. There are many complications such as the effect on oil flow of oil grooves in the bearing, the leakage of oil by viscous flow itself. Most of these have been resolved satisfactorily by the combined attack of theoreticians and practising engineers. There is, however, one area which still challenges both of these groups of tribologists. Under certain conditions, particularly at high speeds, the flow of the lubricant between journal and bearing may become turbulent and unstable. Under other conditions the journal may develop vibrations as mentioned above. In small machinery these effects are usually unimportant. But in the large generators used in modern power stations such occurrences can be disastrous and their economic consequences crippling. The suppression of instabilities in large bearings is a major challenge.

Summary and Prospect 165

Another problem area concerns that type of lubrication which is called elasto-hydrodynamic. This is particularly important in lubricated systems where the contact pressures are very high, for example, between gear teeth or between cams and tappets in an automobile. The high pressures deform the surfaces elastically but in addition they increase the viscosity of the oil. It is this viscosity increase which prevents the oil from being squeezed out and so protects the surfaces even under severe loading conditions. This is now well understood. We do not, however, fully understand how the elasto-hydrodynamic film ever breaks down, for the harder the film is pressed the more difficult it should be to extrude it. Consequently the problem remains as to why these films ever fail. If we can explain this type of failure we may be better able to avoid it. We shall also need more factual data describing the effects of high pressures and temperatures on the flow properties of many different types of lubricant.

Another important aspect of lubricant failure between sliding surfaces is the following. If we add a small quantity of a "surface-active" material (for example, a fatty acid) to a lubricant, it may have very little effect on the bulk viscosity of the oil or on the viscous properties of the thin film trapped between the moving surfaces. Nevertheless it can have a marked effect in mitigating the breakdown of the hydrodynamic or elasto-hydrodynamic film. This is the regime of boundary lubrication. Part of its action is well understood in a descriptive sense. But we still know very little of the strength properties of the boundary film and how these are influenced by its molecular structure. An associated problem is that concerned with those additives which are called extreme pressure additives. These materials usually contain sulphur, phosphorus or chlorine and are dissolved in small quantities in the "base" lubricating oil. If the lubricant film fails and the asperities rub on one another with the danger of incipient seizure the high local temperatures will cause chemical reaction with the additive, thus forming a protective film precisely where it is required. Yet here again we know very little of the detailed chemistry of the surface films or of the most desirable physical properties that these films should possess.

Most of the lubricants used in practice are mineral oils

which have been refined or treated in various ways by the petroleum industry. Others (far more expensive) are mainly synthetic. But both natural and synthetic oils are nearly all composed of carbon, hydrogen and oxygen. Such materials have certain natural limitations; the most serious of these is their chemical stability when their temperature is raised. Particularly in the presence of air they are likely to oxidize; this makes them acidic and likely to attack the sliding surfaces. At a later stage the oxidized products will "cross-link" and form large complex molecules which are very viscous and gumlike. This is the dark brown sludge we find after we have run our automobiles without a change of oil. Yet a further change may occur later and the oil may form lacquerlike deposits on any solid surfaces with which it comes into contact. The chemical stability of oils can be increased by using anti-oxidation additives (oxidation inhibitors) but there is a limit to this. Furthermore, similar types of breakdown can occur even in the absence of oxygen, so that anti-oxidants are irrelevant. The oils and additives used in modern automobiles are surprisingly effective and a tribute to the persistence and efforts of the petroleum industry. But a modern aircraft with its turbojets and ancillary equipment makes far more serious demands on the lubricant and indeed on the hydraulic fluids used to activate many of its mechanisms. There is a great challenge here for the development of new lubricants and new hydraulic fluids with greatly improved temperature stability. It is possible that we may see a move away from carbon-based oils for this very specialized purpose.

Dirt and Seals

Finally we may consider two other problems that are crucial to the practical operation of machinery and are yet concerned neither with the nature of the sliding surfaces nor with the type of lubricant used. These are dirt and seals. Many wear problems and many causes of failure in a running machine are due to the presence of dirt in the oil film. In some cases the dirt enters the system from outside. For example, the hard dust particles which an automobile sucks into its carburetor and hence into the cylinders amount to about a quarter of

Summary and Prospect

a pound per hour of running on a normal road. Consequently we need devices to filter the air. Again wear particles generated by the rubbing process itself may move from one part of a machine to a far more vulnerable part. Clearly we need to circulate the oil and filter it. Filters are indeed widely and successfully used in these and in many other applications. Yet we do not really know how they function, for it has long been observed that they can trap particles far smaller than the size of the pores in the filter material.

The second field of ignorance concerns the action of seals, particularly rubbing seals which separate a running part of a machine from a stationary part. For example, seals are used on the journals of pumps to retain the fluid or gas that is being pumped; on the propeller shafts of ships to prevent the leakage of seawater into the ship; while in large industrial works seals are used around the main bearings to restrict and if possible to prevent the entry of dust and dirt. The seal designer can construct seals that will work successfully over a wide range of conditions provided the machine designer leaves sufficient room for them. Unfortunately in spite of this degree of success the reliability of seals is still a serious problem. It arises from the fact that we do not really understand the mechanism by which the oil film that separates the seal and the shaft is formed and maintained. Further fundamental work here could, if successful, be of very great technological value.

The Prospect

In facing the tribological problems of the future the technologist in industry and the scientist in the laboratory will have many difficulties to overcome. One that they share is the difficulty of communication. They do not understand each other's language and the scientist has by no means been faultless in this regard. The engineer and designer are continuously faced with the problem of interpreting fundamental scientific studies into terms which they can apply to practical situations. The scientist must make an effort to present his work in a way that provides a bridge to the practicing engineer. But ultimately it will rest with the practical men to provide the practical solutions. And part of the responsibility will lie with the attitude of industrial management.

The task of the scientist is more clearly defined. He will need the tools and the techniques of his profession. He will need to use optical microscopy, optical interference, electrical measurements and other techniques to determine the area of contact. He will need to use electron microscopy and the electron probe to study the way in which the surfaces are deformed during sliding and how transfer from one surface to the other occurs. He will need to use electron diffraction to study surface structure and the composition and structure of surface films. He will need to use radioactive tracers to study wear and transfer. He will need to apply mass spectroscopy to a study of surface reactions. And in addition to all this he must match the technique he uses to the scale of the problem he is investigating.

He must also use all the physical and chemical concepts that can be applied to the explanation of his observations. He must know how solids deform elastically, how plastic flow occurs, how contact stresses can produce slip in ductile solids and cracking in brittle solids. He will need to know how surfaces react with environments and how surface films are formed, and what are their mechanical properties. He will need to know how structure and composition affect the strength of new materials under various environmental conditions. And in addition he will need all the experience and guidance that the intelligent and perceptive engineer can contribute to the subject.

It is clear that tribology involves an interdisciplinary exercise. It demands the activities of the physicist, the chemist, the metallurgist, the material scientist and the mathematician as well as the engineer and the designer. In tackling the tasks ahead the tribologist will need elaborate equipment and extensive research facilities. But he will also need scientific insight and maturity of judgment. For however much the development of tribology may rely on instruments, machines, test rigs and computers, its future progress will ultimately depend on men and ideas.

INDEX

Abrasive wear, 153-54
Academies of Science, 12, 13-14, 16-17
Additives, 113, 118 ff. (*see also* specific kinds); and lubricants, 113, 118 ff., 165, 166; and wear, 151, 158, 159, 165, 166
Adhesion, and friction, 8, 16, 18, 23-24, 61-63, 64 ff.: of metals, 61-76; of non-metals, 77-94 *passim;* and rolling friction, 136-37, 138-46; under extreme conditions, 95-107 *passim;* and wear, 148-50, 152, 155, 163
Adhesive wear, 148-50, 152, 155, 163
Adhesives, practical, 67
Aerodynamic lubrication, 113-14
Agricola, Georgius (Georg Bauer), 117
Air, 95, 97-98, 99 (*see also* Oxygen); and lubrication, 113-14, 129; and wear, 151, 152
Aircraft, and friction, 3, 75, 103-4 ff.: braking and, 144; and lubricants, 166; and wear, 92-93, 154
Alcohols, and boundary lubricants, 119, 120, 122

Alloys. *See under* Metals, friction of
Amontons, Guillaume, 13-14, 18, 63, 85, 109
Amorphous structure of solids, 36-38, 40
Angstroms, size of atoms measured in, 27
Anti-oxidants, 126, 166
Aquaplaning effect, 139
Archard, J. F., 149-50
Area of contact, surface roughnesses and friction and, 23-24, 47 ff., 54 ff., 77-94 *passim*, 95-107 *passim*, 147-55
Aristotle, 10, 131
Asbestos, 143
Asperities, 55, 62, 63 (*see also* Surface roughnesses, and friction); and friction of non-metals, 77-78; and lubricants, 124; and wear, 153
Atack, Douglas, 145
Atoms (atomic structure), 25-27 ff., 35 ff.; attractive forces between, 26-27 ff., 44, 61-62; and boundary lubrication, 119, 120, 124; brittleness and, 33; crystals and, 34-35, 37; elastic deformation and, 29-30; fracture and, 33; in free surfaces, 35-36; in gaseous state, 45-46;

Atoms (*cont'd*)
  liquid state and, 42–45; plastic deformation and, 30–32; in plastics and polymers, 40–42; in rubber, 38–40; size of, 27–29
Attractive forces, atoms and molecules and, 26–27 ff., 44, 61–62; long- and short-range, 26–27
Automobiles, 2, 3–4; axles, 74–75, 110 ff., 131; bearings, 75; brakes, 2, 143–44; clutches, 144; door locks, 2; engines, 44–45, 75, 116, 161, 165, 166–67; lubricants and, 116, 165, 166; pistons, 161, 165, 166–67; tires, 3–4, 138–43; and wear, 161, 165, 166–67
Autoradiography, 64, 122
Axles, 74–75, 110 ff., 131

Babbitt, Isaac (and "babbits"), 157–58
Back transfer, and wear, 148, 149
Bakelite, 42
Ball bearings, 73, 132–34 ff. (*see also* Bearings); and wear, 158
Balsa wood, 42
Bearing alloys, 74–75, 76, 158–59
Bearings, 2, 11, 44–45, 73, 74–75; alloys, 74–75, 76, 158–59; ball, 73, 132–34 ff., 158; dry, 163; lubricants and, 74–75, 110 ff., 164–66; rolling, 131 ff., 153; use of PTFE on, 82; use of wood in, 87; and wear, 153, 155, 158–59 ff.
Bentonite, 126
Bismuth, 104
Block gauges, wringing of, 67
Borides, 105
Boron carbide, 105, 106
Boundary lubrication, 117–25, 159, 164; action of, 124–25; breakdown of, 122–23
Brakes (braking), 2, 3, 143–44; drums, 3; materials used in, 143–44; spragging in, 71n; tires and skidding and, 139–43
Brittle solids, deformation and friction of, 33, 87–88, 168. *See also* specific kinds
Bronze, 158, 159
Brush analyzer, 48–50

Cam-and-tappet mechanism, lubrication and, 116, 117, 165
Carbides, 92–94, 102–3, 152 (*see also* specific kinds); and friction at high temperatures, 105–7
Carbide tools, and friction, 102–3
Carbon, 166
Carbon atoms and molecules, 40, 119, 120; in rubber, 38–40
Carbon brushes, 92–94
Cellulose fibers, in wood, 42, 85, 86, 145–46
Chloride films, and lubrication, 127
Clutches, braking and, 144
Coatings, and wear, 164
Coefficient of friction, 7, 12, 19–20, 97, 99, 106, 107; and adhesion, 67, 73; and boundary lubrication, 122, 123; and lubricants, 112–13, 122, 123, 127; and non-metals, 77, 78, 80, 81, 82, 84, 85, 86, 89, 91; rolling, 131–32, 138, 140, 143, 144; and sliding bodies, some representative values of, 7; and wear, 147–48, 150, 162
Cohesion, Desaguliers on, 16
Cold welding, 62
Compressors, and wear, 161–62
Copper, 34, 36, 76, 96, 104; and bearing alloys, 75, 76;

Index

Copper (*cont'd*)
and lubricants, 121, 122, 123; static friction value for, 76; surface atoms and, 36, 43, 44; and wear, 148, 149–50, 158, 159
Copper-lead bearing alloys, 75, 76
Coulomb, Charles Augustin, 8, 16–20, 22, 63, 109
Course of Experimental Philosophy, A (Desaguliers), 14–16
Creep effect, 68
Crook, Alfred William, 116
Cross-links, 38, 39, 40, 41–42
Crystals (crystalline structure of solids), 31, 32, 34–35, 37, 38, 40, 41; and liquids, 42–43; and surface contours, 47–48
Cugnot, Captain, and friction research, 13
Cylinders, 3, 13. *See also* Pistons (piston rings)

Dacron, 40
Deformation, 8, 22–24, 57, 62, 65, 68, 75, 79, 85, 95, 134–46, 168; atoms and molecules and, 27, 29 ff.; elastic, 29–30, 38–40, 41–42, 55, 65, 70, 84, 87, 114–17, 135, 136–37, 138–46, 168; and lubricants, 114–17; of non-metals, 29, 83–85, 87–88; plastic, 30–32, 55, 65, 87–88, 134, 135, 152, 168; in plastics and polymers, 41–42; rolling friction and, 134–37, 138–46, 168; of rubber, 38–40, 83–85; viscoelastic, 87; and viscosity in liquids, 44, 87; and wear, 152, 168; of wood, 85
Desaguliers, John T., 14–16, 20, 23, 61, 65
Devitrification, in glass, 37
Dexrin, 42
Diamonds, 33, 35, 163; and adhesion, 163; and brittleness, 33; crystalline structure of, 35; friction of, 88–89, 98–99, 102
Dirt, role in wear of, 154–55, 163, 166–67
Dislocations, crystalline structure of solids and, 35
Door locks, and friction, 2
Ductility, 32, 98, 99, 106, 168; defined, 32
Dust, and wear, 154, 166–67

Economic factors, tribology and, 160–62
Elastic deformation, 29–30, 55, 65, 70; and lubricants, 114–17; in plastics and polymers, 41–42; in rubber, 38–40
Elastic hysteresis, 40, 135, 136, 138–39 ff.
Elasto-hydrodynamic lubrication, 114–17, 118, 159, 165
Electrical force, attraction between atoms and molecules and, 26–27 ff., 34n
Electron diffraction, method, 52–53, 168
Electron microscopy, 52–54, 168. *See also* Microscopes (microscopy)
Electrons, 34, 52–54, 168. *See also* Electron microscopy
Energy (energy loss), 3, 8, 20–23 (*see also* Deformation; Hysteresis); elastic and plastic deformation and, 29–32; and overcoming friction, 3, 20–23; rolling friction and, 136, 138–46; rubber and, 39–40; "thermal," and liquid state, 43–44
Engineering: nature of metallic surfaces used in, 57–59; and tribology, 156–68 (*see also* Tribology)
Engines, and friction, 13, 44–45 ff., 163 (*see also* specific aspects, kinds, parts, prob-

Engines (cont'd)
lems): airplane (see Aircraft); automobile, 44–45, 75, 116, 161, 165, 166–67; bearing alloys, 75; and wear, 154, 157–68 passim
Environment, effect on wear of, 151, 168
EP (extreme pressure) additives, 126, 127–29, 151, 158, 159, 165

"Fade," braking and friction and, 144
Fatigue, surface, 153
Fatty acids, 165; and boundary lubrication, 119, 120, 121–22, 124, 127, 128
Filters, 154–55, 167
Fluon, 78, 81–82, 90–91
Force-separation curve, atoms and molecules and, 27–32, 33
Fracture, 33, 35, 42
France, early studies of friction in, 13–24, 16–17, 109
Free surface (atoms), 35–36
French Academy of Science, 12, 13–14, 16–17, 109
Friction: adhesion and, 14–16, 61–76 (see also Adhesion, and friction); causes of, 7–8; defined, 4–5; desirability of, 3–4; in everyday life, 1–8; in history, 9–24; laws of, 4–7 (see also Laws of friction); lubrication and, 109–29 (see also Lubrication [lubricants]); measurement of, 4–7; of metals, 61–76; of nonmetals, 77–94; nuisance of, 2–3; solid surfaces and, 47–59; static and kinetic, 67–68 ff.; and surface contact, 23–24, 47 ff., 54 ff., 77–94 passim, 95–107 passim, 147–55; tribology and, 8 (see also Tribology); under extreme conditions, 95–107; wear and, 147–55, 157–68; what it is, 1–2
Frictional heating, 100–7, 144, 145; and wear, 152, 153

Garnets, 89
Gases (vapors), 45–46 (see also specific kinds); cleaning of metal surfaces and, 106–7; as lubricants, 113–14; thermal energy and pressure and, 45–46
Gauges, block, wringing of, 67
Gears, 45; lubricants and, 165; wear and, 160
Generators, 2
Geometry, effect on friction of load and, 5–7, 9 ff., 79–81, 85, 96
Giles, Cyril George, 141
Glass, 33, 37, 38, 48, 139–41
Glues, 67
Gold, 62, 76, 97–98, 99; plastic deformation and, 31–32; wear and, 149–50, 163
Graphite, 91–94
Gravity, friction and loss of energy and, 20–23
Greases, 2, 109, 110 ff., 125–26. See also Lubrication (lubricants); specific aspects, kinds
Grinding, and wear, 153
Grooving, and rolling friction, 134–36
Grubin, A. N., 116
Gyroscopes, 2

"Hard metals," 105–7
Hardness, metallic, effect on friction of, 72–73, 105–7
Harrison, James, 86
Heat, frictional, 100–7, 144, 145, 152, 153
Heating (see also Temperature, and friction): caused by friction, 3, 100–7; frictional, 3, 100–7, 144, 145, 152, 153; and lubrication, 120–26, 127;

Index

Heating (cont'd)
  and surface cleaning, 96–97, 100 ff.; and wear, 152, 153
Heron of Alexandria, 109
Hirn, Gustave Adolphe, 109
Hydrodynamic lubrication, 110–13, 115, 118, 159, 164, 165
Hydrogen atoms and molecules, 26, 40; in rubber, 38–40
Hydrogen sulphide (H_2S), 106–7
Hysteresis, 40, 135, 136, 138 ff.

Ice (and snow), friction of, 89–91; skiing and, 103–4
Indium, 66, 73
Industrial Revolution, 12–13
Industry, and tribology, 159–60 ff.
Interfacial adhesion, 8 (see also Adhesion, and friction); and friction of non-metals, 77–94; and friction under extreme conditions, 95–107 passim; and plowing in friction, 61–63, 64 ff.; and rolling friction, 138
Intermittent motion, static and kinetic friction and, 68–71
Internal combustion engines (see also Engines, and friction): materials and wear in, 163
"Internal" friction, 136
Iodine, active, and lubrication, 129
Ionic bond, 27
Iron, 34, 36, 76

James, Henry, 97
Jost Report, 159–60
Journals (and shafts), 110 ff., 132; and lubrication, 110 ff., 164; and wear, 160
Junction growth, 97–98; overcoming, 97 ff.
Junctions, friction, 63–64 ff., 87–88, 96–97 (see also Junction growth; Shearing): evidence for, 63–64; and wear, 148–50, 151, 162–63

Kinetic friction, 6, 67–68; and static friction, 67–68 ff.; and stick-slip motion, 68–71
Knots, friction and effectiveness of, 4

Lamellar solids. See Layerlike (lamellar) solids, friction of
Latex, 38–40
Laws of friction, 4–7, 13–14, 18, 62–63; first, 6, 13–14, 20; in history, 11; how to cheat, 73–74; second, 6, 20
Layerlike (lamellar) solids, friction of, 91–94, 106–7, 128–29: lubrication and, 94, 128–29
Lead, 66, 75, 157–58, 159, 163
Lead-base materials, and wear, 157–58, 159, 163
Leonardo da Vinci, 8, 9, 11–12, 13, 77, 95
Leslie, John, 22–23
Lignin, 42, 85, 145
Lignum vitae, 42, 86–87
Liquids, 42–45 (see also specific kinds); and gases, 45–46; and reduction of friction, 109, 110 ff., 124, 129, 164, 165; and solids, 42–45; some properties of, 42 ff.; surface tension and, 36
Load, and friction, 5–7, 9 ff.: effect on friction of geometry and, 79–81, 85, 96; and wear, 148
Lubrication (lubricants), 2, 8, 73–75, 109–29, 164–66 (see also specific aspects, kinds); additives and, 43, 118 ff., 165, 166; aerodynamic, 113–14; boundary, 117–25, 159, 165; elasto-hydrodynamic, 114–17, 118, 159, 165; extreme-pressure, 126, 127–29; and

Lubrication (cont'd)
friction at very high temperatures, 104–7; in history, 9–10; hydrodynamic, 110–13, 115, 118, 119, 164, 165; lamellar solids and, 94, 128–29; polymers and, 79; viscosity and, 44–45, 110–12, 116–17 ff., 129, 164

Lucretius, quoted on wear, 147

Machinery, and friction, 2–3, 4–7, 12–13, 164–66 (see also specific aspects, kinds): lubrication and, 44–45, 164–66; wear and, 147–55, 157–68

Machining (milling), effect on friction of, 102–3

Magnesium, 34

Materials, sliding: and lubrication, 164–66 ff.; and wear, 162–64 ff.

Melamine, 42

Metallurgy, and wear, 157–68

Metals, friction of, 61–76: adhesion and, 61–76, 79–81, 87–88; alloys (bearing alloys), 74–75, 76, 158–59; brittle solids and, 33; cheating laws of friction in, 73–74; crystalline structure and, 34–35; effect of hardness on, 72–73; effect of load and geometry in polymers and, 79–81; elastic and plastic deformation and, 29–32; fracture and, 33; frictional process in, 62–63 ff.; "hard" metals and, 105–7; indentation hardness and, 56; lubrication and, 109–29; metallic bond, 26; nature of surfaces used in engineering, 57–59; and rolling friction, 131 ff.; static friction values for, 76; surface contours and contact and, 47–48 ff.; surface oxide films and, 97 ff.

Metal soaps, 124, 125–26, 127–28

Mica, structure and friction of, 35, 48, 53, 55n, 85, 93

Microprobe microscopy, 54, 168

Microscopes (microscopy): electron, 52–54, 168; microprobe method, 54, 168; optical, 51–52, 168; reflection electron method, 52–53, 168; "resolving power" of, 51; transmission electron method, 52

Mineral oils, 115–18, 120, 124, 125–26, 129, 165–66

Minerals, crystals in, 35

Molecules, 25–27, 38–40, 41–42 (see also Atoms [atomic structure]); in gases, 45–46; in liquids, 42–45; lubricants and, 119, 120, 123, 124, 126; in solids, 25–27, 38–40, 41–42

Molybdenum disilicide ($MoSi_2$), 106–7

Molybdenum disulphide (MoS_2), 91–94, 106

Motion, stick-slip, 68–71

Newcomen, Thomas, 13

Newton, Sir Isaac, 11, 12, 14, 25, 69n

Non-metals, friction of, 77–94: brittle solids and, 87–88; diamonds (and other hard solids), 88–89; ice and snow, 89–91; layerlike (lamellar) solids, 91–94, 106–7, 128–29; mica, 85; polymers and plastics, 77–82; rubber, 83–85; teflon and fluon, 78, 81–82; wood, 85–87

Non-stick saucepans, 81

Nuts and bolts, friction of, 4

Nylon, 27, 40, 41, 77, 78

Oblique (or taper) sectioning, 50–51, 62, 63–64

Index

Oils, 2, 9, 10, 110–17 ff., 124–25, 126–29, 164–66; base, 127–28; dirt and seals, 166–67; and greases, 126 (*see also* Greases); and lubrication, 110–17 ff., 124–25, 126–29; viscosity of, 44, 45
Optical microscopy, 51–52, 168
Oxide films, surface contact and, 56, 57–59, 64, 66, 73–75, 76, 95 ff.: and lubrication, 109–29 *passim;* removal of, 95–107; and wear, 152, 154, 155, 163, 165–66
Oxygen, 25–26, 36, 37, 95, 97, 129, 166 (*see also* Air); and wear process, 151, 152

Papin, Denis, 13
Paraffin oil, 127, 128
Parent, M., 13
Petroleum oils, 117, 166
Pistons (piston rings), 3, 13, 154, 161
Plastic(s), 27, 40–42, 77–82; effect of load and geometry on, 79–81
Plastic deformation, 30–32, 33, 55, 65, 134, 152, 168. See *also* Deformation
Platinum, 97–98, 120, 121
Plexiglas (Perspex), 40, 77, 78, 80, 90
Pliny the Elder, 125–26
Polishing, 47–48, 57–58, 102
Polyethylene, 78, 148
Polymers, and friction, 40–42, 44, 48, 63, 66–67, 77–82, 159 (*see also* specific kinds): adhesion and, 66–67, 68; effect of load and geometry on, 79–81; structure of, 40–42
Polystyrene, 77, 78
Polythene, 27, 40, 41, 77, 78; and wear, 149–50
Pressure, and friction, 159 (*see also* Load, and friction; specific kinds, problems): adhesion of metals and, 61–76;
and gaseous state, 45–46; and solid surface contact, 55–57; viscosity in liquids and, 44–45; viscosity in lubricants and, 110–17 ff., 129
Profilometer, 48–50
PTFE, 78, 81–82, 90–91, 147
Pulping of wood, 145–46
PVC (Polyvinyl chloride), 38, 40, 78

Radioactive tracers, 64, 121–22, 168
Reflection electron microscopy, 53–54
Refractory solids, 104–7
Repulsive forces, atoms and molecules and, 26–27 ff. See *also* Attractive forces, atoms and molecules and
Resiliency, in rubber, 39–40
Reynolds, Osborne, 110–12, 164
Rods, metal, 73–74
Rolling bearings, 131 ff. (*see also* Bearings); and wear, 153, 155, 158–59 ff.
Rolling (rotary) friction, 1, 2–3, 7, 131–46; brakes and, 143–45; source of, 134–38; tires and, 139–43
Roughnesses. See Surface roughnesses, and friction
Royal Society of London, 12
Rubber, and friction, 27n, 38–40, 63, 83–85, 155: and lubricants, 114, 115; rolling friction and, 134–36, 138; tires, 3–4, 138–43
Rubbing, and surface cleaning, 98–99

Sabey, Barbara E., 141
St. Petersburg Royal Academy, 12
Scuffing process, 163
Seals, 126, 155, 166–67
Shafts (and journals), 110 ff.,

Shafts (cont'd)
132; lubricants and, 110 ff.; wear and, 160
Shakespeare, William, 157
Shearing, 31, 73; friction and adhesion of metals and, 62–64; friction junctions and (see Junctions, friction); friction of non-metals and, 78; and lubricants, 124–25, 127; rolling friction and, 138; and wear, 148–50
Silica, 126
Silicon, 37
Silver, 31–32, 76
Sintering, 66, 93, 144
Skis (skiing), 82, 90, 103–4
Sliding (sliding friction), 1, 2–3, 5–6, 7, 8; and deformation (see Deformation); and energy loss, 20–23; and lubricants, 109–29 passim, 164–66 ff.; of metals, 68–71, 76; of non-metals, 77–94 passim; and rolling friction, 131, 138–39, 145–46; and stick-slip motion, 68–71; surface roughnesses and, 19–20; under extreme conditions, 95–107 passim; and wear, 149, 155, 162–64 ff.
SNORT equipment, 103–4
Snow (or ice), friction of, 89–91; sliding and, 103–4
"Soaps," metal, 124, 125–26, 127–28
Solids, 8, 25–45; amorphous structure of, 36–38, 40; atoms and molecules and, 25 ff.; brittle, 33, 87–88, 168; crystalline structure of, 34–35, 37; elastic and plastic deformation and, 29–32, 33; fracture of, 33, 35; free surface in, 35–36; friction and contact between surfaces of, 47–59; friction of metals (see Metals, friction of); friction of non-metals, 77–94; and

friction under extreme conditions, 95–107; and liquid state, 42–45; plastics and polymers, 40–42; refractory, 104–7; and rolling friction, 131–46; rubber, 38–40; some properties of, 25–45; surface tension and, 36; wear and, 147–55, 157–61; wood, structure of, 42
Spalling process, 153
Speed(s): effect on wear of, 152; high, friction at, 100–2 ff.
Spragging, 71
Static friction, 5, 6, 67–68; kinetic friction and, 67–68 ff.; and stick-slip motion, 68–71; typical values for metals of, 76
Steam engines, 13. See also Engines
Stearic acid, and boundary lubrication, 120–21, 122, 123, 124
Steel (steel surfaces), 73, 76, 77, 79, 82, 86, 96, 102, 104; crystal structure of, 35; elastic and plastic deformation in, 29–32; and rolling friction, 134–36, 139, 140; surface roughnesses in, 49, 50; and wear, 148, 149–50, 152, 157–58
Stick-slip motion, 68–71
Stylus method of surface examination, 48–50
Sulphide films, 127
Sulphur atoms, 38
Surface contact, and friction, 23–24, 47 ff., 54 ff., 77–94 passim, 95–107 passim, 147–55
Surface contours, and friction, importance of, 47–48 ff.
Surface fatigue, 153
Surface oxide films. See Oxide films, surface contact and

Index

Surface roughnesses, and friction, 14–15, 16–23, 47–59: and adhesion, 61–76 (*see also* Adhesion, and friction; Asperities); and area of contact, 23–24, 47 ff., 54 ff., 77–94 *passim*, 95–107 *passim*, 147–55; electron microscopy and, 51–52; oblique or taper section examination method of, 50–51; optical microscopy and, 51–52; shape of, 54; stylus method examination of, 48–50; and wear, 151, 153

Surfaces, and friction, 14–15, 16–23, 47–59 (*see also* Friction; Surface contact, and friction; Surface roughnesses, and friction); adhesion and (*see* Adhesion, and friction): lubrication and, 73–75, 109–29 *passim*, 164–66

Surface tension, 36

Talysurf Profilometer, 48–50
Taper (or oblique) sectioning, 50–51, 62, 63–64
Tappets, lubricants and, 116, 117, 165
Teflon, 66–67, 78, 81–82, 90–91; and wear, 150, 163
Temperature, and friction, 3, 39, 40, 96–97, 99, 100–7, 152 (*see also* Heating): and gaseous state, 45–46; and liquid state, 42–44; and lubrication, 120–25, 126, 127, 165–66; and surface cleaning, 96–97, 99, 100–7
Themistius, 1, 6
Thermal energy: and gaseous state, 45–46; and liquid state, 43–44
Thermocouple effect, 100–1
Tilted-pad bearings, 112
Tin, 75, 157–58
Tin-base materials, 157–58
Tires, and friction, 3–4, 138–43

Titanium carbide, 102–3
Titanium compounds, 93, 102–3, 106, 107, 129
Titanium disulphide (TiS_2), 93
Titanium iodide, 129
Tower, Beauchamp, 110, 111
Transfer, and wear, 148, 149
Transmission electron microscopy, 52–53
Tribology, 7–8, 157–68 (*see also* Friction); defined, 8; dirt and seals and, 154–55, 163, 166–67; economics and, 160–62 ff.; industry and, 159–60 ff.; lubrication and, 164–66; materials and, 162–64 ff.
Tubes, metal, 73–74
Tungsten carbide, 33, 102–3, 107; and wear, 159, 163
Tungsten compounds, 33, 102–3, 107, 159, 163
Turbines, 2, 45

Vacuum, high, and surface cleaning, 96–97, 98–99
Varlo, Charles, 132–34
Visco-elastic materials, 41
Viscosity (viscous flow), 44–45, 164; and pressure and lubrication, 44–45, 110–12, 116–17 ff., 129, 164

Water, as a lubricant, 139n, 141–42
Water film, and friction, 89–91, 95, 103
Waxes (and resins), in wood, 42, 86, 87
Wear, 3, 8, 147–55, 157–68; abrasive, 153–54; adhesive, 148–50; complex nature of, 155; dirt and, 154–55, 163, 166–67; effect of environment on, 151; effect of speed on, 152; lubrication and, 118 (*see also* Lubrication [lubricants]); mild and severe, 150–51; non-existent laws of,

Wear (cont'd)
147–48; surface fatigue and, 153
Welding, cold, 62
Wheel, the, 1, 4, 13, 14
White metal bearing alloys, 75, 76, 158–59
Wood, and friction, 42, 85–87: pulping and, 145–46; structure of, 42, 48, 85, 86, 145–46; waxes and resins and, 42, 86, 87
"Work-hardening," 32

Yarns, 4

Zinc, 34, 99